John T. Schmidt's Beat:

From Barnyards To Boardrooms

John T. Schmidt
Chancellor Bag, Standard, Alberta, Canada T0J 3G0

Sylvia O'Callaghan-Brown
#1119-942 Yonge Street
Toronto, Ontario M4W 3S8

John T. Schmidt's Beat: From Barnyards to Boardrooms

ISBN 0-9732532-0-7

Cover Artwork by Norah Travis

Printed in Canada by The Ayr News, Box 1173, Ayr, Ontario N0B 1E0

TABLE OF CONTENTS

WORDS OF WISDOM
"The more I learn about agriculture, the crazier it makes me."
Tom Isotamm, Drumbo, ON

ACKNOWLEDGEMENTS

It is a formality and an onus in a book to have a page dedicated to acknowledgements. John T. Schmidt said this and an index are the only two assignments he couldn't write.

It would require 25 pages to list every name mentioned in this manuscript in an index. The economics were bad.

Somewhat the same difficulty emerged in trying to acknowledge the assistance and encouragement given to this member of the Canadian Farm Writers Federation to commission an authorized biography.

There is one exception. That is special thanks to the Ontario Agricultural College of the University of Guelph. Picking the brains of the faculty led him into this small group of farm writers. He became articulate to the point of annoyance on the literary scene.

The O.A.C. has established a farm writers archive to lay before the public the accomplishments of this group dedicated to insatiable stewardship.

Disclaimer: The bible of newsrooms is the style book. All writers are required to use it religiously for consistent capitalization, spelling, hyphenization, titles, length of paragraphs, etc. However, every time a new editor shows up, the first task is to change the style book.Schmidt has been victimized by 27 style book changes; and these changes have been reflected in the text of this book.

ABOUT THE AUTHOR-ANNOTATOR . . .

Sylvia O'Callaghan-Brown was born in Forchheim, Germany, daughter of Maximillian and Anna Starzyk. Her first Canadian home was among the sugar beet farms around Picture Butte, Alta.

A widely diverse work experience took her to all the Prairie Provinces and the Northwest Territories. Her academic career has elevated her to master of philosophy classes at the University of Guelph in the later 1990s.

Her resumé has noted work in the field of psychiatric nursing, the federal income tax department, the oil patch, marketing, management and retail sales and a franchise, banking, finance and money management (the latter in Bermuda).

A change in career took her into the service of Canadian Broadcasting Corporation radio in Yellowknife, N.T., and in Regina to the television newsroom.

A keen interest in political activism was capped by a stint in the office of Premier Grant Devine of Saskatchewan.

In the academic field, O'Callaghan-Brown also earned bachelor of arts degrees in psychology and human justice at the University of Regina. The latter discipline allowed her to go on an international practicum in Atlanta and Beaufort, Ga.

In the field of community work in Medicine Hat, Alta., she served as president of the Canadian Club, a United Way board member, the public library and a writers group.

In the writing field, besides this book, she has been involved as co-author of "Atomic Bombshell," published in 2002, and an unpublished manuscript, "Where Have All The Farmers Gone — Who's Next?" (2003). This has brought her full circle to the agricultural scene, although from the perspective of Toronto.

INTRODUCTION

Discovering the secret behind a long, healthy and extraordinarily productive life is the motivation for compiling this remarkable collection of articles written by a man named John T. Schmidt. For some, the "T" not only stands for Thomas but also for the trouble that distinguishes him as an equal opportunity offender of remarkable quality—occasionally referred to as BullSchmidt. Others recognize him as The Exchequer of Chancellor, which is the name of the Alberta hamlet where he lives and where the residents don't stand on pomp and circumstance. To his more serious readers, however, he is acknowledged to be the incomparable Dean of Agriculture.

No, this is definitely not yet another self-help book in the conventional sense, but it does, nevertheless, examine the ingredients that have gone into making one very good life. And, in so doing, it hopes to provide information useful to others for getting more out of living. This book explores the work of a writer born in 1923, who has been writing since 1938, and shows no sign of stopping any time soon. This examination of Schmidt's writing turns on two fundamental questions:

1. What keeps a person doing what he has been doing for more than sixty years? Long after he could be leisurely checking out concert halls or taking long naps in the warm sun, Schmidt relentlessly pounds out a seemingly endless stream of words. While most of us dream of the day when we can stop working in order that we might do the things that we would really like to do, Schmidt continues doing what he has done for more than six decades: he writes.
2. What keeps people reading the work of a writer who has written more than 10,000 newspaper and magazine columns and several books? Writing primarily through an agricultural perspective, Schmidt's boundless humour, far reaching insights, provocative critiques and enjoyable witticisms continue to provide fresh sparks to ignite his readers into thinking about the issues that affect all Canadians. Looking through the lens of humanity's primary industry, Schmidt sheds light on the interconnectedness of the whole world and the interrelationship of all the people in it.

A watchful observer, he diligently chronicles the events that shape the world in which we all participate. Reflecting Schmidt's writing style, Chapter I, *The Country Messenger*, intertwines biographical information about his personal life and family with the things that are most important to him—communicating with people, humour, global events, his profession and agriculture. Each of the succeeding chapters consists of carefully chosen articles that elaborate his views on each of these things.

Chapter II, *Cross-Fertilizing Ideas*, focuses on the importance that Schmidt attaches to corresponding with others. A consummate man of letters, he strongly advocates that everyone express concerns to one another as well as the authorities rather than resorting to violence or mob action. While he is a staunch individualist, his brand of individualism holds personal responsibility to and for the community as the highest ideal.

Chapter III, *Bucolic Biomass Material*, and Chapter V, *Farmers Frank*, are a collection of vintage farm-writing humour that are guaranteed to split the reader's sides. Some of the people mentioned may no longer be household names, but their types are unmistakably enduring and as such continue to be drop-dead funny.

Chapter IV, *Peripatetic Traveler Through the Gardens of the World*, raises public awareness of other countries and shows the more glamorous aspect of agricultural reporting not usually associated with farm writers. Demonstrating his tremendous breadth of knowledge, Schmidt does not merely report events that occur in other countries, he puts them into an understandable context for his Canadian readers.

Chapter V, *Farmers Frank*, indicates with the coming of *Frank* magazine that Schmidt's writing style drew more attention in the seats of power. Yes, he admits he plagiarized that magazine's style which often got the powerful off their seats to deny everything.

Chapter VI, *Food for Thinking*, traces the changes that have occurred in the Canadian media with respect to farm writing over the course of Schmidt's long tenure in the newspaper industry. He began his career as a farm writer in agriculture's heyday and has witnessed both evolve into very sophisticated operations that he is rightly proud to be associated with.

Through his articles, we discover what the real Schmidt is in his ideas, in his work, and in the contribution he makes to the world. And equally importantly, we find him in the delightful humour he joyfully shares.

In 1991, at the height of his career, John T. Schmidt was presented with an honorary membership in the Canadian Farm Writers Federation at Montreal in recognition of his "long and valuable contribution to journalism". The venue took him full circle to an area where his maternal ancestors were pioneer farmers seven generations ago. Tendering the award was George Price, a top CBC farm writer from Ottawa.

CHAPTER ONE

Country Messenger

THE LIFE OF JOHN T. SCHMIDT

"Curmudgeon" is how John Thomas Schmidt laughing describes himself both in and out of print. Ironically, from the time he first began writing columns for his father's newspaper at the age of fifteen, he always makes sure that the person he criticizes is always left "with a leg to stand on." His self-description, however, not only knocks the props from under himself, but it doesn't measure up to Schmidt's unusually perceptive abilities to make accurate assessments. That isn't to say that he hasn't made questionable observations over the years, but few as obviously incorrect that as the one he makes about himself. The dictionary defines curmudgeon as: a crusty, ill-tempered and, usually, old man. While he has now lived long enough to qualify for the latter, the other two parts of the definition really don't fit. Yes, occasionally he does display an ill-temper when he lets loose oral or solid brickbats. However, these occasions can seldom, if ever, be regarded as anything but righteous indignation. Schmidt possesses a keen sense of justice and has little patience for anyone or any system that mistreats individuals. Far from being "crusty," it is Schmidt's unfailing humour that has presented him with more difficulties than anything else. In fact, the only two instances that resulted in a week's unpaid rest from work were due to his sense of humour that was evidently not shared by the management of the Calgary Herald.

While Schmidt's relationship with management had some bumps, his enduring success as a farm writer can be attributed to the enviable rapport he enjoys with his readership. In the more than sixty years that he has been in print, he periodically reveals a bit of himself and his extraordinary life through his columns. These articles, together with some that have been written about him by others, illuminate his lifelong love affair for providing information that encourages thinking and interesting discussions among his readers.

One of the hallmarks of Schmidt's ingenious writing style is found in his ability to blend seemingly dissimilar topics into informative articles designed to entertain as well inform his readers. He relishes every opportunity to pull ideas together into new and creative ways of thinking about virtually anything, including information about his own personal life. In describing his birthday celebrations in 1988, he cleverly ties some of the world events that occurred when he was born to the day that this:

FARM WRITER WAS COWED BY STRETCHED LIMO RIDE
— Vulcan Advocate – June 15, 1988

The biggest agricultural news that occurred in the month I was born was that the British had removed an embargo on beef cattle from Canada. The month was April, the year was 1923 and the date of my arrival on this earth was the 25th. For the first two years I lived in the second storey of the building, which housed the *Ayr News*, then a small Ontario weekly paper owned by my father, John A. Schmidt, and I haven't been far away from newspapers for the ensuing 65 years. It was a rather humble beginning. My father and his brother, Alfred Schmidt (Uncle Mooney) bought the struggling paper 10 years prior for $500. The business included the upstairs living quarters. When I was two, we moved into a house at the west end of the village next door to a small farm. In addition to becoming a printer's devil, I had the opportunity of learning to drive horses, milk cows by hand, hoe turnips, slop hogs, shovel gravel, and how to plow and pitch sheaves.

All this adolescent activity gave me a hands-on, and often odoriferous, understanding of farm practices—and eventually led to a career as a farm writer. I'm not sure why there was an embargo on beef cattle being shipped to traditional markets in Great Britain in 1923. I suppose Mackenzie King, then Prime Minister of Canada, had something to do with removal of the embargo, however. I've heard numerous old-timers talk about how they plied their way back and forth across the Atlantic escorting boatloads of live cattle to markets there during the 1920s and 1930s.

Very little beef is sold to Great Britain now. In fact, it more often goes in reverse. In recent years Canadian cattlemen have had to seek federal government help to keep from being inundated with manufacturing beef from Ireland. In the years since the Second World War, Canadian cattlemen—especially those in Alberta who have operated on a North American beef economy—have had their share of struggles keeping the U.S. border open. Sometimes it has been the result of non-tariff barriers put up by the U.S. Today, the threat is still there also from Canadians who oppose free trade with the U.S.

The nine million people of Canada in 1923 were paying $478 for a Ford car. But if any of the nation's 25 million in 1988 want a Ford they must pay $17,000. This is nearly half the $38,000 average annual wage—but then the price in 1923 was nearly half the $1,090 annual wage. Gasoline cost about 70 cents a gallon then, compared to today's 45 cents a litre price. Bread was 24 cents a loaf but is now $1.68—but the price of wheat is about the same for both years. In 1923 there was nothing as ostentatious as stretched limousines; probably the only one in daily use was owned by Al Capone. However, with big business and big government and such T.V. programs as "Dynasty" and the "A Team," they are much more common today.

This farm writer happened to be back in his native Ayr on his 65th birthday trying to find a linotype around the *Ayr News*—but found the last one had been scrapped a year before. All typesetting is now computerized. Disconsolately contemplating this reversal in technology, the family decided to raise my spirits that evening with a small birthday bash in nearby Paris, Ontario. Stepping outside my son's house, I was confronted by a shiny, white stretched limousine complete with plush seats, bar, radio, television, telephone and chauffeur. The chauffeur proved to be a local farmer's son who was in the farm implement business but found there was enough limousine business to buy two of them.

The second big surprise was being whisked in the opposite direction to Waterloo to the restored and renamed Kent Hotel. It has been restored to the period of more than a century ago when it and a brewery in a basement cavern were run by a great-uncle, Chris Huether. The third surprise of the evening was seeing a number of familiar looking faces sitting at one of the long rows of tables in the quaint old Huether Hotel tavern quaffing some of the product of the brewery from the vats in the cavern. When my eyes got focused the group turned out to be a dozen friends from my days as a farm writer at the *Kitchener-Waterloo Record*. The group included the Horse Doctor, one of the best turnip hoers in Ontario in his youth in Ayr but an even better city hall reporter and pianist later.

Yes, it was an event worth waiting 65 years for. The most titillating aspect for this barefoot printer's devil and farm boy was riding inside that stretched limousine and imagining the remarks from other motorists, such as:

"I wonder where those gangsters are going?"

"Why, you dirty war-mongering, bloated capitalists!"

"Is the Fat Man making a movie in Waterloo? Where's Jake?"

"Only farmers can afford to ride in style like that."

Thanks, Cec, Bill, and Inez. It was good to have brother Jim, sister-in-law Lavina, daughter-in-law Diane and granddaughter Sonja along for the ride.

Not unlike most fathers, Schmidt's pride and joy are the children he had with his first wife, Jean, who is now deceased. Deceased too is their son, Tom, who died in a car crash at the age of 17. Schmidt married Margaret in 1962 and they live in Chancellor, Alberta. His chest swells with admiration for his three children. John P. (P. for Patrick, having been born on St. Patrick's Day) who is also known as Cec, is married to Diane and the father of Sonja, is the editor of the Ayr News. Bill (William), an engineer in Vancouver, is married to Diana and they have twin boys, Maximillian and Alexander, who keep their daughter, Victoria, company. Inez, who works for Nike, lives with her children, Gabe and Elizabeth live in Portland, Oregon. Brother Jas. (aka Jim) and Lavina live in Ayr, where he rides herd on the newspaper that has been a family operation for four generations. Schmidt credits his Uncle Mooney and Hughie Melvin, a linotype operator at the Ayr News, for encouraging creativity in his writing and his father for instilling him with a sense of responsibility to the community for what he writes in his columns. In his heart he also has an indelible soft spot for the Women's Institute which played a large role in both the history of Canada and in Schmidt's personal development as a reporter.

W.I. DISCOVERS COLUMN
– *The Calgary Herald* – August 20, 1970

In an August edition of *Canadian Magazine*, my friend and former Herald staffer, David Cobb, gave a commendable run-down on the Women's Institutes and the influence for good they have had on the rural communities across Canada. Cobb quotes Dr. Ed Ballantyne, Deputy Minister of Agriculture for Alberta, as saying to Mrs. Marian Alexander of Cayley, President of the Alberta Women's Institutes: "You stand for wholesomeness, positiveness and the goodness of our rural community. I often wonder what would happen if you weren't there." For the edification of Mr. Cobb and Dr. Ballantyne, this column wouldn't have been written today had the Women's Institutes not been there. I spoke at a Civic Holiday celebration in the east-central Alberta village of Cereal—at which the Cereal W.I. was honoured for its publication of the village history, *Down Cereal's Memory Trails*—how the W.I. first made me aware I had a product that would sell. I would like to give further acknowledgment here.

My family published the weekly newspaper in Ayr, Ontario. So we wouldn't be hanging around Wong's poolroom, stealing hockey sticks from a bankrupt factory, or hitchhiking across the country in our bare feet, my brothers, Jim and Robert, and I were early indoctrinated into what we considered the drudgery of work in the newspaper. When my dad discovered my aptitude for sweeping floors and carrying water was almost zero, as a last resort before notifying the reform school, he gave me the only thing left to do: write a column. I was 14 at the time.

The column consisted of bits and pieces that wouldn't fit in between the ads or anywhere else. We had a linotype operator, Hughie Melvin, a taciturn gent with a dry wit who got off wisecracks or off-colour stories several times a day. I used a great number of these, albeit dry-cleaned a bit. I'm not sure whether it was one of his stories or a smutty poem that was brought up on the floor of the Women's Institute meeting the following week but there was terrible outrage over it. They may even have passed a motion of censure. They discussed a few other things off the record that day too—socio-economic matters like: whatever was going to happen to that pregnant hired girl on the fifth concession, the terrible brawl at the Queen's Hotel Saturday night, the miracle that Mrs. Smedge's husband was speaking to her again, and when were they going to get rid of Peck Barnard from the general store, that nasty old man?

But the item in the column was on the record—and one of the neighbours took the trouble to report that fact to my dad. She added that if I continued to debauch and poison the minds the women and children of the village with the kind of suggestive pieces I was writing, I would end up by being drummed out of the village choir. I received orders to clean up the column or I would be put back to sweeping floors again. That was my first lesson in journalism. I figured if the column was being read by members of the Women's Institute it was being read by everyone—and hence the encouragement to keep at the business for many more years.

At Cereal, in addition to acknowledging the debt of gratitude to the followers of Adelaide Hunter Hoodless for directing my energies towards writing, I was also happy to acknowledge the contribution to the written word of this nation the Women's Institutes have made through their Tweedsmuir histories. To my mind, the feel of the real history of rural Canada is contained in these home-brewed works. History as written by the professional historians can be deadly dull and stuffy and full of dates and places that never bring out any human interest. And what is history really other than a great number of human experiences told by the people who were actually on the scene at the time?

When official recognition is given for this contribution, I would like to think the efforts of Cereal W.I. will be near the top for all of Canada. What pleases me most of all is that this great contribution to history by Cereal, and hundreds of other Women's Institutes, has been made by women of the rural community. They led the way. All great human endeavor has its genesis in the rural areas. As long as there are community spirited groups, like the Women's Institute in Cereal, to exert a unifying force on the community this gives the community the hope of remaining viable in the Canadian fabric.

The difference between the two approaches was concisely illustrated on my departure from Cereal that warm Saturday night. It was only natural for the convivial crowd in town that night to demolish a few brews. This brought two members of the Oyen detachment of the RCMP loping into Cereal to dampen boisterous spirits by the use of a roadblock near the railroad station. I dutifully stopped at the red light and a young constable shone a flashlight in my face. Deciding I wasn't one of "that element," he motioned me on—at which point I took the opportunity of reviving one of the bon mots of the old linotype operator and took off down the main street lined with people.

Suddenly, in a great flurry dust and red lights, the squad car pulled me over and two stern young constables ran up and accused me of: 1. Lack of proper respect for the RCMP and 2. Drinking. Just as they had started the "busting" procedures, they were confronted by the presence of my wife, who had been indoctrinated into the devoutly righteous world of Women's Institutes at an early age. "Leave my husband alone, you young whipper-snappers," came her comment at the end of a short, sharp exchange of words. The RCMP hightailed it, with apologies. Whoever heard of the RCMP retreating from the Women's Liberation Movement?

Five generations ago, a strong widow's passionate concern for liberating her family from the incessant warfare in Germany compelled her to buy her three sons out the army to dispatch them to Canada. Her sentiments have run deep in most of the Schmidt family ever since. Individual responsibility for the family and the community is a serious commitment for the Schmidts.

Being a precocious lad, Schmidt was accelerated twice in elementary school and definitely he had a mind of his own. As might be expected, this did not always permit him to see things the way his father did. Having inherited his ancestor's strong will, he demonstrated his independent spirit by turning his hand to farming. The years spent working on a farm proved to be extremely valuable when he returned to the newspaper business. He worked as a linotype operator, photographer and reporter at several different newspapers before he settled into the farm editor's chair at the Kitchener-Waterloo Record for seven years prior to heading out for Calgary.

Schmidt's work for the Kitchener-Waterloo Record was not without its glitches but the downside was always put back into perspective when he received mail from his readers who appreciated his efforts.

THE LAST COLUMN
– The Kitchener-Waterloo Record – *February 1, 1958*

We received a letter from Mrs. Harold Lamoureux of Preston, Ontario, who said:

> "It was good to see a comment in your column on the remarks made by two McMaster professors that they could see no real reason for even attempting to save the fruit-lands of Niagara. Their statement seems evidence of a shocking blindness to the true needs of men, and left me feeling rather hopeless about the future. If such a remark had been made by a businessman, we could see why he would be thinking thus; but

where can we look for wise leadership when university professors show such dangerous ignorance? Can you further write on our critical need to look beyond the next shareholders' dividends? My husband and I find in your writings a sincere endeavour to look below the surface of things and present the truth, however unpopular that may be at the time."

This excerpt from Mrs. Lamoureux's letter has pleased us immensely for it indicates that she has perceived what we have been up to for the last seven years. When we write "30" at the end of this column it will mean the end of our association with this newspaper as farm editor. We leave next week to take a position with the *Calgary Herald.* So it doesn't look as if we will get around to belabouring the "dangerous ignorance" of those who, even from pioneer days, have had a preoccupation for cutting down trees. One way or another we have tried to make it our business to learn something new about agriculture and its allied fields every day. Some of the things we learned did not seem highly praiseworthy and we often said as much. On the other hand, we were pleased to note genuine progress and noted likewise. Some of our criticism at times didn't go down too well with leaders of one organization and another as they considered criticism a personal affront.

The one justification we can offer is that any organization without criticism dies of apathy and indifference. If anyone doubts this just look at the tobacco, hog and peach marketing boards sporting officers with large salaries and see how vital and active they are. The reason is that they are constantly kept on their toes by snipers. Writing a mish-mash of pleasing platitudes was never our stock in trade, we are happy to say. While one tends to grow cynical in this business, we find much gratification in discovering that the thinking of people is always changing. A rule of thumb five years ago cannot be accepted today.

It is sometimes too easy to become impatient with what one considers the bungling and foibles of the human beings entrusted to run our affairs. We know at times we were like Mrs. Lamoureux; we did feel hopeless about the future at times when we felt that leadership suffered. However, this feeling is of only momentary concern to us because in a young country like Canada if there is any feeling of hopelessness for the future it is because people do not realize the full potential of the nation. Speaking of age, we have met numerous people who were acquainted with our writings before they met us. Invariably they made the comment: "I expected to see an older man." We can't understand that. At the age of 34 we would hate to be known regularly as an old curmudgeon.

The last seven years have been ones of hard work and have lent a goodly amount of satisfaction. We were born in this part of the country and it is good to see that it is one of the fast-growing and richest areas of the nation. To some degree, it may become dehumanized by the pressure of living as we know it here. Our desire to live in the West has been impelled by quite a few trips there in the last couple of years. We hope to gain the same appreciation of it as we have of this area.

Having spotted a conservative bent to Schmidt's thinking, the editor of the Calgary Herald offered him a staff position. In the fifties the West was an exciting place for agriculture and Schmidt was anxious to learn more about what was happening there. So off to the West went the young man, where he still lives today.

The first big difference he noticed, when he first arrived, was that the attitude of the Western people seemed to be more free and easy under the Big Sky. Farm organizations and individuals in Ontario took criticism very personally. Westerners, on the other hand, seemed to him to be less contentious. Agriculture in the West consisted primarily of cattle ranchers and grain farmers and their controversies made for interesting copy. In the East, the conflicts were mostly about marketing boards, which eventually spread to the West but their approach was quite different.

Recognizing that the Calgary Herald served both an urban as well as a rural audience, Schmidt promises, in his first Herald column, to keep both informed about the exciting events in the food production industry. For Schmidt, this promise was a solemn oath that he unfailingly upholds to this very day.

FIRST COLUMN - *The Calgary Herald* - September 2, 1958

This is by way of introduction to a new by-line which will be appearing over this column for some time. The last Agricultural Alberta column was headed "Farewell—For a While." In it, Tommy Primrose explained that his doctor had ordered him to take a leave of absence due to ill health. It is good to report to Tommy's many friends who have asked about him that he is taking a rest at his mother's home at High River. When I talked to him on the phone Monday he said he was "coming along fine." It is possibly only coincident with this good report that Tommy has bought a ranch at High River. He says it is all of one acre but there is a little stock to wrangle and a few things to fix up. I couldn't think of a better way for a man to get back on his feet again than spend a few hours outdoors like this every day.

Now a word about me:

For seven years I sat at the farm editor's desk of the Kitchener-Waterloo Record looking over the Ontario farm scene. In the early 1950s this was a relatively peaceful scene but when I left in February of 1958, it had become a scene embroiled with wrangling over methods of marketing livestock and crops. The marketing of hogs was causing the biggest trouble—and I understand that battle is still in and out of court. Agriculture in the East is not essentially different from that in the West. A few different methods may be used and the Western farmer may operate on more acres than his Eastern counterpart but both are in business to make a livelihood by any means at their disposal.

Many persons in the city do not realize the extent to which science and technology have changed the farm scene. Many farmers themselves do not realize the extent to which they may be aided by new developments in science and technology. Farm organizations have been providing a great deal of spot news and will continue to do so as farmers fight the battle against the cost-price squeeze. These are only a few of the doings on the farm scene, which make it exciting and newsworthy these days. They will provide the basis of commentary for many forthcoming columns.

It is my aim not only to try to keep farmers up-to-date on matters of vital interest to them but to keep city people informed about Canada's basic industry. The reason I believe that city people want to be—and should be—better informed about developments on the farm is that many of them have a farm background and still retain an interest in farming. Then, too, developments in technology have made it possible for one farmer to produce 10 or 15 times as much as his counterpart of a quarter century ago. This has caused a drop-off in farm population and this drop-off has reached the point where the farmer now belongs to a minority group and, as such, his interests must be constantly championed before the masses who depend upon him for subsistence.

There are a certain number of "giants" in every industry whose words are regarded as gospel and for whom a reporter is pleased to be among the number who "sit at his feet." One of the men whose words a reporter could depend upon to hit the headlines is Professor Ralph Campbell of the economics department of the Ontario Agricultural College in Guelph. He does not look like a professor. He would walk into a meeting looking as if he had come in fresh from sowing a wheat field or milking the cows. When he started talking the air would become electric.

This year he is President of the Agricultural Institute of Canada. Some farmers regard him as too socialistic and radical. Many regard a recent speech of his along the same line, for he said that the mechanical revolution that farmers have seen in the last 25 years will be nothing compared to the revolution in the next 10 years, especially in the production of hogs, broilers, eggs and turkeys.

The revolution will bring a further drop-off in farm population, continued overproduction and the related need for price supports and other government aid and adjustment for farm people in marginal areas to enable them to keep up with the changes. Revolutions—bloody or bloodless—always bring problems greater than experienced to date. The big problem in the old days used to be how to harvest a crop fast so that its nutrient value would not be lost. The advent of bigger and faster machinery enabled the farmers to virtually lick this problem. The limitations to success in the raising of field crops are poor management, bad weather and bugs.

Professor Campbell says these limits don't apply with the same force in the raising of hogs, chickens and turkeys. Some of the rankest amateurs have entered these enterprises and become immensely successful in them—possibly because they were not bound by tradition. He is of the opinion that there is no limit to the expansion potential of the enterprises specializing in these fields.

The only thing that Schmidt likes better than writing letters is to write them on hotel stationery purloined on one of his countless trips both here in Canada and abroad. The following application to the fictitious University of Gumbo in Chancellor, Alberta, written on the rather distinctive Tokyo Prince Hotel stationery, is a good example of Schmidt's clever use of humor to defuse any questions about his creditability for writing a column without having a degree in agriculture or journalism.

In all seriousness, however, Schmidt's journalistic education began as a teenager, writing columns for his family's weekly newspaper and later he went on to write for the dailies as well contributing numerous articles to various magazines. Working on a farm gave him the first hand experience to develop invaluable insights into complexities of farm problems. As a devoted listener to CBC Radio, he gained a good background in the Arts and doing research for the books he has published enabled him to acquire a substantial understanding of history. Taken in combination, Schmidt epitomizes the self-educated man who equips himself well to meet the challenges of the world.

APPLICATION TO ENTER THE UNIVERSITY OF GUMBO – CHANCELLOR, ALBERTA

As will be noted from the application, my institutional education was confined to high school. I didn't go to university as it would have been a waste of time at that point. None of the left-wing radicals and theorists from the United States and other countries had arrived in Canada when I was ready to sit at their feet. They were all underground during the Second World War.

I fell into agricultural journalism by accident, having been given a six-month trial at the *Kitchener-Waterloo Record*, which lasted for nearly seven years. At the end of that time I moved to Calgary because agriculture was more or less drying up in the industrialized part of Ontario and remained semi-dormant until consumers discovered eggs were costing them more money as the result of the Canadian Egg Marketing Agency destroying two million dozen to hold up the price. The West was the place where the agricultural action was and is and I jumped into the middle of it at the *Calgary Herald*. When I first started writing in agriculture about 70% was confined to covering the activities of rural groups and indicating to farmers how to improve their husbandry techniques. This has completely changed today. Today 70% of agricultural writing is involved with politics as governments have inserted themselves into agriculture to a greater degree either imagined or thought possible by the public at large.

I feel that a smattering of formal political science and economics at university would better prepare me to understand why government depends to a great extent on professionals from these fields for their advice and excursions in agriculture. These excursions have generally placed agriculture in a terrible position, with governments shoring up the holes with more and more ad hoc legislation and regulations. I feel that eight weeks' exposure to a campus would put me in touch with the thinking of younger people, most of whom at present seem to feel that I have no sympathy with the revolution, pot smoking, shooting up dope, women's liberation, Maoism, obtaining government grants, multiculturalism, and liberal attitudes to damn near everything, including prison reform. I feel that my ability to foment violent public disagreement in my writing, plus the fact that I have now grown a beard, would allow me to fit into campus activities without too much dissension.

One of key ingredients to Schmidt's writing technique is to listen to classical music while he works. His earliest memories include those rare but euphoric occasions when he had finally saved enough money to take the train into Toronto to attend the symphony. Among CBC's most dedicated listeners, he has seldom missed a broadcast in the past sixty years of Texaco's Saturday Afternoon at the Opera.

Over time, he developed numerous and innovative ways of enticing and sustaining the interest of his readers, many of which will be discussed in detail in other sections of this book. Intuitively, Schmidt is an imaginative writer who is quick to spot fresh ideas and takes great delight in passing them with his readers.

AGRICULTURAL EDITOR EXPLAINS SECRETS OF HIS WORK
The Calgary Herald - November 3, 1959

Not a few persons have asked this writer why he has not used his picture on the masthead of this column. That's a good question and I think the answer may be found in the above picture. I have a high degree of compassion for the sensibilities of the readers.

Another question frequently asked is: "How and where do you gather enough information to write a column five days week?" Without giving any trade secrets away I can reveal that there is a barrel sitting in one corner of the office. Beside it sits the rubber boots seen in the photo. I try to throw as many ideas in the barrel as I take out. However, at times more ideas are taken out than go in. And when I get to scraping the bottom of the barrel and find nothing there, I just dip into the old rubber boot. So far this method has never failed.

I am very grateful to some of my old pals on the *Kitchener-Waterloo Record* for sending along the rubber boots, pitchfork and overalls which I had left there. They were accompanied by a telegram which expressed the relief of the staff that "the air is clear" again. For the benefit of my Kitchener friends—and every dyed-in-the-wool Albertan will confirm this—everything on the Western farm scene is not only bigger and better, but deeper and thicker. I may soon need a stepladder.

To keep this column going and to keep Ken Liddell happy, I intend to carry on further research into the reason why a milk stool only has three legs. I have my own ideas but they may not bear the light of scientific scrutiny. From one source I learned that much early furniture, including tables and milking stools, was built with only three legs. Cabinetmakers later learned that furniture was more stable with four legs. Yak yak!

Other research has revealed that from a fellow named J. Haskin Flint of Montpelier, Vermont, you can buy a do-it-yourself milk stool kit. Mr. Flint has capitalized on the fact that most of the cows today are milked in the milking parlour. No more of this sitting on a three-legged milk stool beside old Brindle hoping she wouldn't kick you into the gutter. So the three-legged race of milk stools has undergone a wonderful transformation and has been relegated to the parlour of the home. My friend from Montpelier peddles his do-it-yourself kits that are "all ready for you to finish to match your decor. Made of Vermont maple, it's fine for T.V. or sitting pretty at a cocktail party," he says. If things keep going this way I'll have to do more research into the horse-laugh.

His extraordinary tenure enabled Schmidt to develop a large network of contacts with whom he shares information and ideas. His contacts help him to keep his readers abreast of the new developments here in Canada and, indeed, all over the world. He travels extensively and has written many columns about agriculture from dozens of countries, some which are included in another section of this book.

Despite having worked for sixty-odd years, Schmidt has yet to master the *difference between working and taking a holiday. He rarely wears a tie but the pocket of the white shirts that he always wears is never without a collection of pens, including a pen watch. Tucked safely behind his pens is a wad of notes written on bits of paper, which make up his own rather unique version of a filing system. Both the scraps of paper and the pens are ever ready to jot down ideas for future columns. In addition to sharing the information acquired in his travels, Schmidt occasionally offers his readers some tips on how they might save a little money.*

LITTLE QUAKE EVIDENCE - ALASKA
The Calgary Herald - December 23, 1964

The extremely cold weather of the past week has no doubt convinced many of my rural friends they should be in California or Hawaii. I know Wilf Edgar of Innisfail will be off to the island paradise by air. Art Berreth of Beiseker assures me that going by boat is the best way to travel to the Pacific isle. For any of the readers from the rural areas who take holidays in winter instead of summer and are looking for a "different" type of vacation, I recommend they shiver their way to the nearest travel agent and sign up for a "Visit USA" ticket. This ticket allows the tourist unlimited travel on 13 regional air lines of the United States for two weeks for $100 plus exchange (a 45-day ticket may also be obtained for $200). Children 2 to 21 can travel for half that fare.

As I like traveling by air this was the greatest travel bargain of the century, although I understand the rates may change February 1, 1965. I can't understand why more Canadians haven't heard about this. The plan is a better deal than a Eurail pass, the widely touted 99 days of bus travel for $99 (which has since been withdrawn) and that new pass on Canadian railway for foreign tourists because it is possible to travel more miles. As with every plan of this type there are a few restrictions but they are not bothersome. I discovered one of these limitations while chatting with an airline clerk in San Francisco International Airport. He indicated that I had only traveled to California using two airlines and that a clause in the ticket would rule it invalid unless I used a third airline. "Well, what other airlines do you have?" I asked.

"Oh, there's Alaska Airlines . . ."

"Go no further," I said, "book me through on a side trip to Fairbanks and Anchorage." Which he proceeded to do with dispatch in late October of 1964.

It was admittedly quite a change in climate—from the 80-degree temperature of Long Beach to the 40-below temperature of Fairbanks—in the space of a day. However, Fairbanks was much more to my liking than the high-priced California rat race. The headlong rush of population to California is apparently more than the state can cope with comfortably. Taxes have gone up out of sight and are reflected in costs all across the board.

During a visit to Alaska about five years ago, I found some of the prices staggering. Prices there are reasonable now compared to those of California. It also goes without saying that freewheeling frontier hospitality and friendliness is an attribute that is becoming less and less noticeable in once languid southern California. It is this spirit that has been responsible for the amazing comeback of the earthquake-damaged city of Anchorage. The only evidence of the major catastrophe of March 27, which can now be detected by a stranger, is a few empty basements where buildings were razed and a few deep holes with trees growing at the bottom. Life goes on as usual. A few people will tell amazing stories about their experiences if pressed. There is an eerie feeling for the outsider walking abroad in the city. The knowledge, that seismographs have registered several hundred quakes since the big one, makes for that feeling. Will another big one come along? The tourist's yen for a bit of far-out adventure is usually satisfied in the North. I got my fill on the return leg: just out of Fairbanks. The Alaska Airlines jet took off and began circling the city. The pilot came on and said he had been unable to retract the right landing gear and would continue to circle. In a few minutes he was back on with the welcome news that the other two wheels had been let down and he had three green lights signifying all were in position. He proceeded to make a normal landing to "check it out." The ground looked pretty good upon landing, although I had visions of hurtling along on one wheel and ending in a heap.

After a ground crew spent an hour applying heat to the sluggish hydraulic system, the jet took off again for Seattle. There was a great deal of banging and crashing. Again the pilot came on with the fearsome news that he again couldn't retract the right landing gear. He gave out with the information that he had radioed the head office at Seattle for instructions before proceeding. He said there was a possibility of hopping over the mountains to Anchorage where the weather was warmer. Just at that point the co-pilot could be heard on the intercom with the welcome news the gear had finally been retracted. The plane lurched ahead to normal speed. The pilot said we were now off to Seattle. With all the gear tucked up the heavy drag on the ascending plane had been relieved—not to mention the feelings of the passengers. I can't remember the pilot's name; he was a good man to have in the cockpit. And, yes, facing a crash, I did see my life whirring before my eyes at supersonic speed.

My travels in the last month in the line of duty and on vacation have taken me over 10,000 miles by every means except horseback but including mail truck from Lillooet to Lytton, B.C. After planes, it's hard to go back to traveling by train, especially on the Canadian Pacific. Those airline stewardesses spoil peripatetic travelers for the manhandling they get from the freight-maulers they call conductors on passenger trains. The conductor on train No. 8 of December 12th had a real good time giving the passengers a working-over. I was so glad to see that trip come to an end that I jumped off at the post office and ran screaming all the way to the *Herald* building.

The only other difficulty I discovered in all those miles was trying to find U.S. airline timetables in Canada. The reason for their absence is sabotage by the customs and excise department. In an idiotic ruling dating from the dark ages, Canadian customs classes timetables as commercial advertising and subjects such material to a duty of 7%. I understand a printers' and lithographers' lobby first pressed for such "protection" for their members. The end result of this unrealistic ruling has meant a hardship on the Canadian traveling public. Many U.S. airlines simply refuse to send timetables into this country for the information of travelers because of the unfair levy. Any bureaucrat who construes a timetable as anything more than information should be banished from the long room. He's no better than Scrooge.

In keeping with his penchant for interconnecting places, people and ideas from all over the world, John and his wife, Margaret, make a point of visiting a town called Ayr in Australia. Visiting places called Ayr appears to be a family trait. His brother, James also organizes trips for his community band to Ayr; but this trip is from the Ayr in Canada to the one Scotland. Delighted with John and Margaret's visit, the newspaper in Ayr, Australia ran a feature column about the differences in climate and lifestyles of the towns called Ayr located on three continents.

VISITOR TO AYR, BORN IN AYR, CANADA
The Advocate – Ayr, North Queensland, Australia – February 23, 1970

Visiting in Ayr for three days last week was Mr. John Schmidt, agricultural editor of the *Calgary Herald* of Calgary, Alberta Province, Canada. Mr. Schmidt comes from a newspapering family and found it interesting to compare notes with the staff members of *The Advocate*. He was born in Ayr, Province of Ontario, Canada, where members of his family have owned the weekly newspaper, *The Ayr News*, since 1913. His father bought the paper in that year when it had a circulation of 500 and over the years it has expanded into nearby villages and now boasts a circulation of 3,000. The paper is presently operated by his brother, Jim Schmidt, with family assistance.

Ayr, Canada is a village of about 1,000. It is located in the industrial heart of Canada 70 miles west of the City of Toronto, capital of the province. It is the home of several small industries but most of the people commute to nearby small industrial cities to work. Agriculture is carried out on small farms of less than 200 acres in size. Pig raising, poultry farming and beef cattle are the main agricultural pursuits. At this particular time of the year, Ayr in Canada is in the midst of the winter season. The temperatures will range down below zero and there will be several feet of snow on the ground. The homes, however, are built substantially of wood, brick or stone and are insulated against the cold. Heating is provided by large furnaces or several small stoves which are kept burning constantly with coal or oil. The cost would be about $300 a year for the average home.

Naturally the people don't go around in shorts as they do in Ayr, Australia. They wear heavy woollen clothing and underclothing in the winter months—that is for about five months of the year. During the other months the climate is more temperate but can reach 90 to 100 degrees in mid-summer in July. During the cold months much rye whisky is drunk to keep them warm. There is only one hotel in Ayr, Canada and that is the Queens.

The village was founded over 150 years ago and was named after Ayr Scotland, as many settlers from Scotland emigrated and cleared farms out of the bush in the pioneering days. Ayr and the nearby town of Paris, Ontario, have an excellent brass band. Last August they chartered a plane and paid a fraternal visit to Ayr, Scotland, that hosted them royally. Their arrangements in Ayr, Scotland were made by Billy Dunlop, editor of *The Ayr Advertiser*. There is a great deal of communication between the two Ayrs. Mr. Schmidt said he has also discovered an Ayr in Nebraska and an Ayr in South Dakota in the United States, which he wants to visit some time soon.

He said he is much impressed by the wide streets here and the visible signs of wealth from the sugar cane and rice crops. Schmidt has been traveling in Australia for a month. He is accompanied by Mrs. Schmidt, who teaches school in the city of Calgary specializing in instruction of emotionally disturbed children. She is on a sabbatical leave this year and is studying teaching methods in Australia. She is acquainted with many Australian teachers who have been lured to Canada—especially Western Canada—by promises of high salaries. Most of them are extremely capable teachers who do a good job. One aspect of the situation is unfortunate. Many have been disappointed by the wages they received when they arrived because they were not given all of the facts by school board trustees, who had recruited them in Australia. Many are sent to isolated areas, which resemble the outback in Australia and some are not prepared for the cold winters, when temperatures go down to 40 below for weeks on end in the West and North parts of the country.

Margaret, too, was much impressed by the warm weather here and the warm reception received in traveling through the great country "Down Under."

Schmidt's family has been in the newspaper business in Canada for more than 140 years on both his mother and father's side. His brother, Jim, took over the Ayr News after their father died and now John P., son of John T., is the editor. Jim underwent a hip replacement operation in 2000, which sent him into semi-retirement. The Ayr News produces a weekly paper through the combined efforts of six family members as well as several other members of the Ayr community.

A strong pillar of the community for all of his life, Jim was honoured by having a park named for him. It is a wonderful ballpark, complete with lights, and it holds great promise for many youngsters in the future. In addition to being one of the chief promoters of the Ayr-Paris Band, Jim is a devoted connoisseur of German cuisine.

JAS. WOULD LIKE THIS CUSTOM-MADE SAUERKRAUT
The Calgary Herald – January 30, 1981

My brother Jas. (Jim) Schmidt, who publishes the *Ayr News* in Ontario every week, attributes much of the paper's success to the fact he has retained a second generation sauerkraut maker. This is Martin Tillch Jr., of Waterloo. His father before him made sauerkraut for Jas. Jas tried it himself many years ago but it never turned out too well. So Martin, Sr., took pity on him and brought in a crockful every fall until his death. Then Martin Jr. took over.

I had tried to ensile cabbage into sauerkraut but somehow it never quite came off, either. Some of the Hutterite people from the Sunshine Colony at Hussar learned of my plight and suggested I appear with a number of cabbages and a crock and they would put their skills to work for my culinary amazement. I duly appeared there one Monday morning with half a dozen cabbages from my garden. They were somewhat the worse for wear, having been viciously attacked by cabbageworms. However, most of the worms had been sorted out. My friends laughed over the fact I had brought a five-gallon crock. I soon learned why.

Elizabeth and her husband, George, and some interested bystanders adjourned to the basement of the colony kitchen and made preparations. They proceeded to make coleslaw out of the cabbage with a special motor-driven cabbage slicer invented by the shop boss. They put the slaw in a stainless steel tub about two feet square. However, in place of a bottom it had fine steel-mesh. The mesh was small enough that the ground-up cabbage wouldn't go through it.

"All right," said George, "take off your shoes and get in and tramp the cabbage." I saw him give a couple of the other folk standing around a sly wink. I figured they were kidding—and did nothing.

George disappeared and came back with a pair of rubber boots. It was easy to trace his previous movements on the boots through a muddy barnyard. Elizabeth carefully washed them off under the hot-water tap. George donned them and jumped into the tub and started to tramp the cabbage. The tub had been raised off the floor on two cement blocks. This was the most important part of the process. It was necessary to squeeze as much water as possible out of the ground cabbage. George tramped it for about 10 minutes until he was satisfied all the water had run out the steel mesh bottom onto the floor.

I was then I realized why they had laughed at the five-gallon crock. The ground cabbage was now reduced in volume to about six quarts. A gallon jar and a half-gallon jar were produced and Elizabeth began cramming the almost-dry cabbage into them after salting it. "The trick is to pack them airtight so the cabbage will cure just like ensilage," said Elizabeth.

If air is allowed to get in, the sauerkraut will turn mouldy and lose its taste. She filled the jars over the top then put a cement block on top of each. It had packed down to the top of the jars by nightfall and the airtight cabbage began to cure. At the end of two weeks it was ready to eat. It had a fine flavor, wasn't too moist and gave off a pleasant aroma. There was no further necessity to keep it airtight, so it was packed in small plastic bags and put into the deep freeze.

I am much pleased to have such professional sauerkraut makers as George and Elizabeth in the neighborhood. I am sure that I will write a better column as a result of their superior product. It goes well with a rasher of bacon. I am also sure it will compare well with the best product Martin Tillich, Jr., prepares for brother Jas.

Among Schmidt's most persistent annoyances are the strange "doings" of the various "gumment" departments. A relentless advocate for the rights of the individual, Schmidt takes every opportunity to expose laws that are passed under the guise of being for the public good but ultimately infringe on personal freedom. While his opinion on various issues may be contentious, there can be no doubt that his views are firmly anchored in individual integrity and personal responsibility. Rather than stepping on the nearest soapbox, Schmidt finds that expressing his views in a humorous way encourages people to discuss important issues while enjoying a good chuckle.

AND THEN THERE IS THE JOHN SCHMIDT OF THE CALGARY HERALD — *1st Column in The Winnipeg Free Press Weekly – April 7, 1975*

What was unacceptable yesterday is acceptable today.

It is thus I have the dubious distinction of being the first schizophrenic to be asked to contribute a column to this estimable farm journal. The trouble is I'm not certain if I should be certified as a farm writer or jailed as a political writer.

When I first started writing a column in the *Ayr Gun* (since renamed the *Ayr News*) in Ontario, I was reporting the best methods for farmers to keep Canada thistles out of the back pasture. Now I'm reporting on methods being used by the Canadian Cattlemen's Association to keep Agriculture Canada out of their pasture fields. To ensure my return to sanity and my desired station in life as a political writer, I have developed the:

Ten New Commandments for Farmers

1. Farmers can do nothing politically for themselves. A strong, charismatic leader is the answer to all problems.
2. Farmers should indulge in economic class warfare and keep pesky city consumers off marketing boards.
3. Each farm organization needs a constitution enabling it to pass resolutions sure to be pigeon-holed by politicians.

4. A large staff of farmers who are too lazy to farm should be hired to see if they will do any work for a farm organization.
5. Farmers should convince themselves that if the government is made up of "baddies," the opposition must be made up of "goodies." (Two elections ago the Alberta Socreds were the "baddies." Now there isn't enough of them to be considered "goodies").
6. Even though they pass NDP legislation, politicians who speak of the "evils of socialism" and "income stabilization assurance" must be "goodies."
7. When things get bad, organize big rallies and invite speakers to tell those farmers assembled things are getting worse.
8. Build an image—not "left" or "right"—but stay in the "middle of the road" (where it's easier to get knocked over).
9. If by chance politicians make election promises and break them after the election, they should be forgiven. Politicians are entitled to their shortcomings the same as everyone else.
10. By observing the first nine commandments farmers can ensure the family farm will soon be a thing of the past. They should plan to sell their farms to the government land bank or get on the nearest government payroll so I, as a political commentator, can write fulsome paeans of praise about their good works.

Having grown up in an era when letter writing was the primary means of communication at any distance, and in a family that publishes a weekly newspaper that crucially depends on the post office, Schmidt expends much effort encouraging better mail service.

TIME IS RIGHT FOR REPLACING POSTAL SYSTEM
The Calgary Herald – July 28, 1982

I have been trying to persuade Harvie Andre, MP for Calgary Centre, to start a movement to cancel the post office and start again. There are compelling reasons for this drastic move. The most pressing is that an almost complete breakdown has developed in reasonable written communications between persons who depend upon the post office. As a person who maintains a part-time residence in Andre's constituency, I am at present barred from receiving communications from him that he sends to the rest of his constituents through the post office. I will say more about that below.

On July 8, 1982 I attended upon the Glenmore Inn where I was advised Broadwater Farm Services Limited and *Agripress Canada Limited* had mounted a seminar on farm land buying and were bringing to Calgary Merrill Oster, president of the Professional Farmers of America. The hotel management informed me that the sponsors had cancelled the seminar. Cancellation, they believed, was the result of plans by the Alberta Department of Agriculture to hold a seminar at the same time. Getting on the phone, I called my farm-writing friend, Julian Bayley of *Agripress* at Hensall, Ontario, to demand an explanation. Bayley had contacted me in advance to inform me the previous day of this seminar and another on how to cope with the economic climate of North American agriculture at the present time. He also asked me to make a note of these seminars in this column. To my embarrassment, I did so on July 3, after they had been cancelled. Through the phone call to Hensall, I learned from Karen Lippincott that a letter had been mailed to me 10 days previously informing me of cancellation. "Are you not aware it takes the post office 10 days to two weeks to move mail from Ontario to Alberta?" I asked—and shuffled through a number of letters on my desk for confirmation of this fact. "No. We use a rule of thumb of four days," she said. Four days in a world of instant communication is no better than 14 days! This is ample reason for the post office to be stamped out and replaced by a 20th century communication method. The economics of replacing the post office system are excellent. In addition to the hundreds of millions of dollars, which must be picked up by the government for losses incurred by the post office, businesses are able to write off equal amounts in federal taxes for phone bills for checking out and keeping written communications up to date. That $10 call I made to Hensall is repeated millions of times a week.

Miss Lippincott confided the seminars had been cancelled because of low registration and bad timing (pre-Calgary Stampede and poor national economic conditions). The land-buying seminar had been held twice in Eastern Canada and once in Regina with success. She thought it would go over in Calgary some time in the fall. She didn't think the registration fees of $175 and $250 had kept away any potential registrants, however.

Earlier notification of cancellation would have enabled me to join the Lacombe Research Station staff in a formal celebration of their 75th anniversary. This is one government service that can be justly proud of the work it has done on behalf of its fellow Canadians. On this date also was a tour sponsored by the Alberta Agricultural Service Boards, which took a wide swing of modern agricultural enterprises in the Municipal District of Foothills centred at High River.

After having said the post office should be cancelled, I will be challenged to defend this position by persons who say this service has become an integral part of their lives, which they cannot do without. In defence of this position, I might indicate that Sam Payne, a post office cop attached to the Alberta postal district, has made me an involuntary guinea pig to demonstrate that it is quite possible to get along without mail delivery. In fact, it is a pleasure to have the burden to playing the post office game lifted from my shoulders.

Payne cut off my residential mail service. Then he turned around and wrote a confidential registered letter to me at the *Herald* April 29, 1982 to inform me of the embargo. The mail from Andre and others is being held at Letter Carrier Depot No.1 in Calgary. As far as I am concerned, it will be held there forever—or until the government scourges Payne and all the rest of them and starts over with new people. The embargo occurred because I tried to impress upon the mail carriers that I did, indeed, live at the address written on letters forwarded to me by my employer. The carrier was returning letters to the *Herald* marked "No Such Suite." This was despite the number being plainly visible on the address.

After having four letters returned with this nonsensical balderdash, I attempted to show the letter carrier there was such a suite by leaving one of the offending envelopes at the mail box and saying in jest "well, come on up and I'll punch you in the nose." This attempt at levity resulted in Payne slapping on his embargo to mail delivered to that box—not the residence, mind you, just my box. The irony of the whole matter must be doubly apparent to the reader.

The amazing part of this piece of infantile acrimony was my delighted discovery that I could get along. There was no expectation of service that would be interrupted by union sanctions gone mad. The only Canadian who is really going to suffer Payne's pique is Harvie Andre, the Postmaster-General. Andre doesn't know how to cope with Payne. He can't write and ask me to vote for him because of Payne's intercession. When this kind of impasse occurs, the only thing that is really going to solve it is to elect MPs who will give permanent maternity leave to the post office and hatch a new plan of communication. Maybe the provinces should take over the mail service and give it back its credibility and sense of dedication and service.

In his public writing as well as in his personal life, Schmidt is not one to quietly accept as "fate" the consequences that result from other people's actions. Strenuously and unrelentingly he objects to the "rules" enacted or enforced by various authorities if he perceives them to be an inconvenience or disadvantage to individuals. A typical example of how a person can get caught in the cross purposes of groups with different interests can be seen in one of Schmidt's personal dilemmas.

DETAILS OF MY ARREST
Memo to Boss at The Calgary Herald - November 14, 1975

I thought I better bring you up to date on being arrested tonight.

On October 9th I received a speeding ticket. On October 19th I pleaded guilty to this offence and as per instruction on the ticket, dropped the ticket and the fine into the post office while in Olds on October 19th in time for it to arrive at the office of the Crown in Calgary by the due date, October 23rd. On October 21st the post office went on strike and the plea of guilty plus the fine are now in the post office. I was unaware of this fact. There was no way I had of knowing it.

Tonight I was sitting at home listening to the opera when two constables walked in and arrested me on a bench warrant for not appearing in court on the charge. I showed them a receipt for the money and a copy of the covering letter I sent with the money and they said I was lying and took me down and threw me in jail. Later they released me on my own recognizance although I had to make an undertaking to appear in court on the same charge on November 28th.

This puts me in the position of facing the court twice on the same charge. I mean they already have the money I sent them as once a letter is posted it is their property and I can't go to the post office and either get the letter out or go to the post office and claim a refund for the money they claim they haven't got. If the post office strike ends before November 28th they will have received their money yet I am forced to go into court and face the charge again. This put me in what is known in law as double jeopardy.

Since I was on a Herald assignment when all this happened I was wondering if I could make use of the *Herald's* lawyer, George Crawford, to see where I stand legally on this matter as I'm sure you don't want on staff a reporter with a criminal charge over his head. I think there may be a good case against the Crown for false arrest, as they have not taken any recognition of the postal strike, and I'm sure they are aware of it. As I understand it, from the arresting officers, there are quite a few incidents of this kind happening because of the postal strike. I figured this was a lark a the time but I am now pretty sore because the cops took my belt away from me and my pants kept falling down. Also the bail man made me pay $4 to get out and said if I wanted to recover the money I could apply to the attorney general's department.

Signed, your law-abiding agricultural editor,
John T. Schmidt

Schmidt's views are most often aligned with the ranchers who for him personify humanity's free and independent spirit. And, periodically, he clearly demonstrates an anti-socialistic bias, but it would be too hasty to conclude that he must, therefore, be an ardent capitalist. A closer reading of his work reveals that, in fact, he is neither. Essentially, Schmidt's overarching philosophy consists of a strong aversion to any kind of monopoly that is detrimental to the interests of people as individuals.

ALLEGIANCE IS TO COMINFORM
– *The Calgary Herald* – June 13, 1972

Already Paul Hellyer, leader of Action Canada has been labelled a "red-baiter" for his criticism of the government for keeping the public in the dark about alleged Communist subversion in Canada. Technically, the critics of Hellyer are right. But they are right only because of a fine technical point. In Canada, the Communist Party of Canada is a legal political party; it has the same status as any other recognized political party and it has run candidates in municipal, provincial and federal elections.

It is not against the law to be a member of the Communist Party of Canada. This is not the case in the United States, where a member of the Communist party is regarded as the agent of a foreign power. The question of whether the charge of red-baiting will stand against Hellyer must be resolved in the context of the status of Canadians who are card-carrying members of the Communist Party of Canada.

The Communist Party of Canada, although legal, is unlike other political parties in that its allegiance is to the Cominform. All the national Communist parties throughout the world are controlled by the Cominform. A good case could thus be made that members of the Communist Party of Canada owe their allegiance to the Cominform and not necessarily to Canada. Each Communist in a democratic country could thus constitute a threat as each has accepted body and soul the Marxist doctrine. The Marxist doctrine contains the seeds for overthrowing democracies.

Any member of the Communist Party of Canada who opposes decisions and orders of the Cominform is quickly purged. In Communist nations such purges are almost always preceded by methods or actions in which the "guilty" person signs a public "confession" of his deviation from the party line. A Canadian Communist either follows the Cominform party line or he is purged. The greatest strength of the Communists is not in large numbers but in blind, unswerving, unquestioning loyalty.

Unlike other Canadian political parties, the Communist Party of Canada uses a standard tactic to fight off criticism and accusations. It seeks to discredit the critics or the accusers by labelling them "red-baiters." Up to now such a label has served it admirably in destroying the credibility of Canadian critics. In fact this writer has been jointly accused of being a red-baiter by the *Canadian Tribune* national organ of the Communist Party of Canada and Roy Atkinson president of the National Farmers Union. The objects of the NFU president's displeasure were apparently two small items that appeared in these columns of October 21 and December 7, 1971. The items quoted James Rawe of Edmonton and Strome, Alberta, on the direction he thought the leadership was taking NFU at its annual meeting in Winnipeg December 6 through 9, 1971.

Mr. Rawe was fired as the NFU's director of organization for Alberta and, as such, had some knowledge of the pedigree of Atkinson's ideas. In a December 15, 1971 front-page article under the by-line of William Beeching, secretary of the Communist Party for Saskatchewan, the *Canadian Tribune* quoted Atkinson as "launching a bold scathing counter-offensive against extreme reaction... Atkinson publicly denounced red-baiting as harmful to the farmers' cause." The *Canadian Tribune* revealed "during the past few years materials have also been circulated all over the country trying to link the NFU with a Communist plot."

Mr. Atkinson is quoted as saying "they are calling us subversives. He told the delegates the impetus for the campaign comes from Communist-haters—'people who are stupid.' He named the Canadian Intelligence Service, Pat Walsh, its research director, and John Schmidt of the *Calgary Herald* as the main source of the anti-Communist campaign against the NFU," the *Tribune* said.

Beeching's tactics here are to use Roy Atkinson to try to discredit a farm writer. The reader will notice, however, neither the *Canadian Tribune* nor Atkinson attacked James Rawe. Commenting on this curious fact, Mr. Rawe said: "I am puzzled as why Atkinson didn't attack me as I have led the campaign against his leadership. The only thing I can figure is that he wants to make it look like the attack is from sources outside the NFU. Had he included my name, many people would have started wondering if there wasn't truth in what was said as I would naturally have inside information. He makes it sound as if the only people linking the NFU with Communism are those of outside interests, rather than including many who were executive members of the NFU. I had nothing but admiration for the NFU as it was originally explained to us. What I opposed were the new progressive policies being undertaken by NFU leadership which are directed at placing of all farmers under total control of a super-elite, thus destroying all rights of the individual to pursue his own initiative and his liberties."

What happened, in actuality, was the *Canadian Tribune* neatly fell into the trap of coming out in the open to attack critics of NFU leadership and at the same time defending that leadership. This brings up the question: Does the NFU leadership need the sort of defense provided by the voice of the Communist Party of Canada? I put this question to James Rawe. His answer was an emulation of Mr. Hellyer's suggestion that possibly Canadians are not fully aware of the lengths to which the Communist Party of Canada have gone to take over this Canadian farm organization.

Human history has been and continues to be profoundly influenced by the powerful weapon of humour. Often employing humour for numerous reasons himself, Schmidt rightly questions the rationale behind the use of humour in the case of an

American Agriculture Secretary. Understanding the underlying motivation behind the publicly stated reasons can at times be quite astonishing.

BUTZ HANDED JOKE BOOK
– *The Calgary Herald* – October 22, 1976

Why would a person with the intelligence of U.S. Agriculture Secretary Earl Butz utter a racial slur about blacks in the presence of a reporter from such a paper as the *Village Voice* only to be promptly fired by President Gerald Ford?

The answer, my friends, is the politics of wheat.

He desperately needed a cover story to get out of his agricultural job while the going was good. That off-color joke about Negroes was guaranteed to do it in the U.S. political climate today. Why did he want out? To allow Ford to slip in some grain price floors again. That story has been played well down in the publicity surrounding Butz's exit.

This isn't the first time Butz has wanted out. Back in June, 1975, a *Los Angeles Times* dispatch to the *Herald* indicated U.S. farmers were after his free-enterprise hide because of falling grain prices. Wheat was at $6 a bushel in February of 1974, and by mid-1975 fell to $3. Fortunately it recovered and Butz's "free market" philosophy won out. However, the price level didn't hold. It slid down slowly at a bad time: during a presidential race. Ford couldn't afford an albatross on his team whose popularity was plummeting to zero among the farm bloc of voters. He couldn't very well hand Butz a pistol and tell him to go out and get more Russian orders for wheat or play Russian roulette. He handed him a joke book and told him to play racist suicide. Which Butz promptly did.

Shortly after his resignation his replacement announced a higher floor price of $2.25 a bushel for wheat and correspondingly higher prices for other grains—over Butz's dead body. With the slide in the price of wheat, the politicians will have the problem of telling consumers why the price of bread won't come down. It's hard to make them believe that since there is only four or five cents worth of wheat in the price of a loaf of bread only a cent or two a loaf could come off. An awful lot of money is waylaid between the farmer and the grocery store—and guess who's getting that money in their pay-packet?

Public sensibilities regarding humor, as well as politics, change over time and occasionally these shifts can be quite dramatic. Schmidt's own sense of humor did not always coincide with those of the Herald's management. It is interesting to note that in the same column, right after the Butz story, Schmidt is able to write an update on the 1/2 Ass Ranch which only a few years earlier had resulted in a week's suspension for him.

1/2 ASS RANCH REVISITED
— The Calgary Herald – October 22, 1976

The first time I came upon Mel Jones' ranch north of Winfield was about a dozen years ago, I nearly swallowed my cigar and my wife averted her eyes at the gate sign: 1/2 Ass Ranch, Mel Jones, Prop. I brought back a photo and ran it on this page and Dick Sanburn, then editor-in-chief, hit the roof at what he considered a piece of barnyard humor on my part. Of course, those were the days before the newspapers had generally started using all those four-letter words that came out of women's meetings and explicit sex stories... everything starts on the farm first!

Last month, while on the annual tour of the Alberta Farm Writers Association, we were entertained by the Winfield Agricultural Society when I ran into Mrs. Doris Jones. She said "yes, the notorious gate sign is still up" and it is still creating a lot of laughs. It was given to her husband, for a birthday present by Stan Panek, his brother-in-law. We had the bus stop as we passed the ranch on the way home and everybody had cameras at the ready. With the distress in the cattle industry today, a great many more farmers think they are running an outfit comparable to Mel Jones' 1/2 Ass Ranch.

With the passing of time, the public was becoming more accustomed to seeing "colourful" words in print, however, the politicians, at least publicly, felt they needed to question the names used for some the ranches.

1/2 ASS RANCH NOT REGISTERED
– *The Calgary Herald* – June 28, 1977

Mrs. Doris Jones of Winfield was perturbed a couple of months ago to read in the *Saint John's Edmonton Report* (later known as *The Alberta Report* and later still, *The Report*) that Gordon Stromberg, MLA for Camrose, had cavilled her family for "having the gall" to register their farm name, which is the 1/2 Ass Ranch. She called a few people including Rusty Zander (MLA for Drayton Valley) and this writer to indicate the Stromberg fascination with the gate sign was misplaced. Stromberg has written to clarify the fact that while he was discussing farm names in the legislature, he had not given honourable mention to the 1/2 Ass Ranch in that august chamber of wisdom and decorum.

What he was doing in the legislature was sponsoring some amendments to the Names of Homes Act, administered by the Alberta Department of Agriculture. The amendments decree that the registrar may refuse an application for a farm name, which has already been requested or a name which the registrar feels is not in good taste or the public interest. These decisions can be appealed to the Minister. Up to this point neither the Registrar, nor the Minister, has had to deal with an application from the Jones family. This is simply because they have not applied for registration of the 1/2 Ass Ranch as a farm name.

There may be many reasons the Joneses have not applied. An increase in the price of beef to a profitable level may make the name instantly obsolete. The MLAs could assist in assuring liquidation of any application by keeping their fingers out of the beef business and telling Agriculture Minister Eugene Whelan of Canada to do likewise. There might be some difficulty concerning the question of taste. For instance, the Registrar might have medical problems that would cause him to crack a rib with mirth if confronted by an application from Jones. He might have to refuse the application on the grounds there is another ranch in the Whitecourt area with the same name. Or he might have to tell the Whitecourt owner to cease and desist if he issued the name officially to another farmer.

Stromberg said that since the legislature passed the act in 1921, 1,389 farm names have been registered in Alberta. Some are extremely original. In 1966, J.R. Wright of Calgary registered Bucki-n'em Palace; Percy Willis of Calgary registered Hell's Half Acre in 1974; E.J. Howe of Caroline officially named his farm Howe-Dee in 1974, and A.L. McNumarer lives at Mortgage Manor at Vegreville. But things aren't all as bad as the last name would indicate. R. Thompson of Vermilion calls his place Optimist and H. E. Sawchuk of Boyle has Prosperity, which nobody else may duplicate officially. One, G. Fox, of Calgary, applied for and received the name, Unto the Hills, in 1961. Edwin John French of Warner had the honor of being the first applicant under the act. His place was named Fair Acres on June 29, 1921. He was followed a few days later by George A. Walker of Viking, who registered the name Laurel Grove.

Incidentally, Stromberg points out that to make things official the Department of Agriculture sends the owner an engraved certificate which can be hung over the fireplace or any other location in the house.

While Schmidt's humour was enjoyed, in fact, encouraged by most of his readers, Herald management once again took a dim view of the following joke and insisted he take on another week's "rest" without pay.

SCHMIDT NEVER FORGETS
– *The Calgary Herald* – July 26, 1978

Jim Miller, co-chairman of the Canadian National Junior Angus Heifer Show, said last time he saw me he told me a story that I promised to run, but that it never appeared so here it is:

The locale was the bankruptcy sale of a well-known Black Angus breeding stud in the U.S. Just as the sale got under way a plump middle-aged women arose and demanded to "kiss the man who is having this sale." To the general applause she kissed the owner, then grabbed a microphone and announced: "I always like to be kissed before I get screwed!"

Miller apparently didn't understand you can't tell barnyard jokes in a family newspaper and Schmidt obviously believed that his readers' taste toward what they saw in print was not that much different than the jokes enjoyed at public meetings. However, the management of the Herald insisted on defining what constituted humour and, more importantly, what information was suitable for public consumption.

Over the years, thousands of farmers left the land to try their hand at various kinds of work in the urban areas, and Schmidt was in a constant battle with Herald management for space to keep the public appraised of these profound changes. Nevertheless, for reasons best known to themselves, Herald management reduced his daily columns to four per week.

LIVESTOCK, CROP MARKETING WOES STILL GOOD COPY
— *The Calgary Herald* – April 19, 1986

The first paragraph of the September 2, 1959 five-day-a-week Agricultural Alberta column read: "This is by way of introduction to a new byline which will be appearing over this column for some time." That "some time" stretched to 27 years. In those 27 years, this farm writer has filled a quota of approximately 6,075 columns with 4.8 million words on agriculture and agribusiness. This quota amounts to some 30 average-length books and constitutes what may be the longest-running five-day-a-week column ever produced by a *Herald* staffer—a fact which gives this high school drop-out some satisfaction. And, incidentally, that word-slinging has made me a lot of money, for which I am thankful.

Every farmer knows the distress of having his quota cut. Now that tribulation has overtaken this column. The quota will be reduced to four columns a week beginning Monday because of editorial space considerations. The readers will also be exposed to it on different days: Monday, Wednesday, Thursday and Saturday from now on. I found it somewhat jovial to read a few paragraphs from that September 2nd column. I outlined the fact I had come from an agricultural scene in Ontario, which during the late 1950s, "had become embroiled with wrangling over methods of marketing livestock and crops. The marketing of hogs was causing the biggest trouble—and that battle is still in and out of courts."

Well, hey, what is new? That same situation in Alberta is grist for a good many columns today. And here are some more of the words from that ancient column: "Many persons in the city do not realize the extent to which science and technology have changed the farm scene. Many farmers themselves do not realize the extent to which they may be aided by new developments in science and technology. Farm organizations have been providing a great deal of spot news and will continue to do so as farmers fight the battle against the cost-price squeeze."

These are only a few of the doings on the farm scene which make it exciting and newsworthy and will provide the basis for commentary for many forthcoming columns. It is my aim not only to try to keep farmers up to date on matters of vital interest to them but to keep city people informed about Canada's basic industry.

More of the same will be written Monday.

The sudden disappearance of his column from the Herald, a few months later, left many readers and fellow journalists wondering about what happened to Schmidt.

Reporter for the Alberta Report Magazine, Owen Roberts, provides some information for those who are/were dismayed or just down right curious about Schmidt's departure.

THE EXCHEQUER CHECKS OUT
– *Alberta Report* – November 10, 1986

After almost 30 years Schmidt leaves the Herald.

John Schmidt, 63, the self-proclaimed "manure editor" of the *Calgary Herald* for nearly 30 years, never did make front page with his quirky daily agriculture report. Nonetheless, the portly journalist has become a barnyard name in southern Alberta thanks to the column. Thus readers were somewhat surprised to learn three weeks ago that *Herald* management had decided that "the need to educate urban readers concerning agriculture... is best done through a large number of different news stories rather than a personal commentary." Neither the employer nor the employee will say if Mr. Schmidt was fired or if he quit. But he's definitely gone.

Other *Herald* reporters are less surprised by the development. The controversial and volatile writer has often annoyed his bosses. The most memorable example occurred on a Sunday night last April after the veteran newsman had just finished his column. Suddenly the company's computer crashed, sending the agricultural composition into alphabet heaven. Mr. Schmidt, for his part, went into orbit, snatched three dinner rolls left over from a staff party the evening before, and hurled them at unamused technicians in the newspaper's computer room. The next day *Herald* brass suspended Mr. Schmidt for two weeks.

There were other tales, too. Fellow journalists recall Mr. Schmidt's penchant for driving *Herald* staff cars until they were stopped by cops, at which point he would abandon them, hitch-hike to his home, and instruct the newspaper to "pick up the keys from the cop." The Canadian Wheat Board, often maligned in his column, would periodically receive correspondence from the columnist written on letterhead inscribed "John Schmidt, Exchequer of Chancellor."

Unorthodox habits aside, Mr. Schmidt was one of the West's most widely read farm writers. With reporting in his blood (his family still owns the *Ayr News*, in a small town outside of Kitchener, Ontario), he worked for the *Kitchener-Waterloo Rec*ord and the *Hamilton Spectator* before joining the *Herald* in 1958 where he become renowned for his crusades against such venerable institutions as the wheat board, and recently, Alberta Agriculture. "He was himself a spokesman for individual farmers," says Alberta Wheat Pool communications manager Douglas Brunton, former agriculture reporter for the *Red Deer Advocate*. "They would phone him, and he'd often take up their cause."

One of his biggest fans was Elaine Deeg of Langdon, 18 miles east of Calgary. Mrs. Deeg heads an eight-member farm women's group called "Decision With Vision," which she says was one of those "little guys" Mr. Schmidt periodically wrote about. She was shocked to learn about his departure. "His column was the first page farmers turned to, to see what was going on." The angry farmwife wrote to *Herald* assistant publisher Kevin Peterson, who wrote back assuring her that the paper "remains committed to agriculture coverage, but because of the wide range of concerns in the industry, we must provide a broad range of coverage." Mr. Schmidt's column, he delicately suggested, was too narrowly defined.

Mr. Peterson claims Mr. Schmidt "left" the *Herald*. Publisher J. Patrick O'Callaghan won't talk about it. "I never discuss staff members with anybody," he sniffs. *Herald* insiders say Mr. Schmidt was fired, a rumour which, on his lawyer's recommendation, the usually outspoken journalist will neither confirm nor deny. He does, however, say he'll write stories for southern Alberta weekly newspapers. But that's not good enough for Mrs. Deeg. "I wish he would have written about the circumstances under which he left. A lot of people are wondering."

Economics and politics when mixed with humour in the public media periodically produce quite a volatile cocktail. As Schmidt observed in his 1976 column about U.S. Agriculture Secretary, Earl Butz, public statements may or may not reflect reality. The official Herald management position maintained that their disagreement was with Schmidt's sense of humour. The end began with the column about one of Schmidt's pet peeves—the Post Office.

'ROYAL FLUSH'—SOLUTION TO POST OFFICE PROBLEM?
— *The Calgary Herald* - September 27, 1986

On Victoria Day holiday, a friend, Ernest Nimitz, celebrated the Queen's birthday by going 50 kilometres into Fort St. John, B.C., to collect the Royal mail from his lock box at the post office there. He was sorely dismayed to find the post office lobby, which is supposed to be open 24 hours a day was, itself, locked.

The great Westerner, the late Senator Harry Hays, enunciated the philosophy of the West that the "latchstring is always out." It was a great shock to my friend to find the post office had yanked in the latchstring, thus reducing postal service to tenth rate from seventh rate. Upon enquiring the reason for the further lapse in service, he was informed by the zone postmaster at Dawson Creek, B.C., that the Fort St. John post office was suffering from vandalism and people fouling the lobby.

As many farmers are forced to do today, my friend decided to retain a consultant: Professor Guy Solonetzin, a social scientist who is president emeritus of the University of Gumbo at its carrigana-covered campus at Chancellor, Alberta. "How," he asked, "can the post office continue to deliver a government program to the people, namely the mail, when faced with a social problem of this magnitude?"

Prof. Solonetzin disappeared down the garden path to a small building with a half-moon cut in the door, located on the perimeter of the campus. He returned some time later hitching up his braces. "You must realize," he said, "that over the last decade the post office has been gutted of human and material resources. It no longer has the depth of professional competence seen in the days of railway mail cars. Exhaustion, obsolescence and savage cost-cutting have run a once-proud capability into the ground to the point that a spiral of disasters such as experienced at Fort St. John can practically close it up."

But there is hope. The post office recently began a social program. At local post offices, millions of dollars were spent inviting clientele in for free coffee and doughnuts as a gesture of goodwill. These tea parties must have missed the persons who are vandalizing the Fort St. John post office box lobby. The post office also built ramps for crippled persons.

And then Prof. Solonetzin continued: "Maybe the post office officials never twigged to the fact there may be a real need among the citizens of Fort St. John to use a washroom—as I understand the post office is near a hotel in the city. Carrying the coffee-and-doughnut program one step further, could the post office not erect washroom facilities (even temporary) in the box lobby to accommodate these poor souls who have been caught short—it may be after a long trip to town for the mail?"

Prof. Solonetzin even had a name for the facility: The Royal Flush. The good professor said he didn't think postal officials would find any trouble drawing upon its social and public relations budget at Ottawa to make the lobby of the Fort St. John post office the first in Canada to be equipped with indoor plumbing. However, if Ottawa can't see its way clear to install the Royal Flush, my friend has offered to mail in a small indoor portable privy from Boondockville for use in the post office of Fort St. John.

No sooner had this version of the column hit the street when some rather offensive material really did hit the proverbial fan. The series of events began with the entrance of Ernest King Nimitz, who was born in Texas near the famous King Ranch. A graduate in range management from Texas A and M University, Nimitz worked in the Alberta Department of Agriculture prior to moving to a bush farm at Sunset Prairie, 24 miles from Fort St. John, British Columbia.

Unable to get any satisfaction from the district postmaster, Nimitz contacted Schmidt, who promptly provided a solution his column. However, Gary Park, the business editor of the Herald at the time, took it upon himself to change Schmidt's intellectual property. Schmidt complained to Park about changing the zone postmaster's words "defecating" and "urinating" to "fouling the lobby." Park, in turn, complained to J. Patrick O'Callaghan, the publisher, about the "abusive" memo he received from the farm writer protesting the continual changing of farm copy without consultation. Park added he would no longer work in the same newsroom as Schmidt.

O'Callaghan called in Schmidt and, making a new instantaneous rule, said: "The words, defecating and urinating, are not used in The Herald. Therefore, you had no right to complain about Park's action in removing these words. You are fired. Be out of here in four minutes."

In a contradictory statement, he then praised the newly jobless farm writer for providing first-class copy in his 28 years on the paper. With this praise ringing in his ears, the confused firee was gone.

O'Callaghan left to take a position in Toronto, where he passed away. In 1998 Park was dismissed after it was allegedly discovered that he had been selling other journalist's articles to trade magazines prior to running them in the Herald. Schmidt, however, continues to write for a number of weeklies and occasionally permits himself to reflect on the acknowledgements he has received for his work by his peers:

FARM WRITERS POINT OUT: YOU CAN'T READ YOUR OWN TOMBSTONE — *The Calgary Herald* - July 26, 1980

Can't read your tombstone when you're dead: The letters are not clear. The Alberta Farm Writers Association, a group of hard-working practitioners, has been held together for the last couple of decades by a policy of never arguing about the constitution when they should be drinking beer. And, a group of the writers was sitting around one night last winter doing just that. Out of that session the writers developed the philosophy seen in the above two lines of doggerel composed by the doyen of the group, Frank Jacobs, Poet-Lower-Yet, to the livestock and farming communities. (I was not present that night as I don't drink beer.)

I am much pleased and delighted to report with as much modesty as I can muster, that I am the first beneficiary of the new philosophy. Accompanied by diverse accolades and a parody of doggerel, the farm writers let it be known to the world that having a wry sense of humour can be a means of surviving more than 30 years on the farm beat on daily papers in Kitchener and Calgary.

In a burst of spontaneous good fellowship the writers struck the first "Can't Read Your Own Tombstone" award. It was presented during a barbecue at, of all places, the Dawson Creek Legion, during the joint B.C.-Alberta farm writers' tour of the Peace River country. The presentation was made by Mrs. Alice Switzer, a special person and farm writer who visits milking parlours and pig sties for the United Farmers of Alberta Co-op. She expressed the hope I would be able to continue for an appropriate period a "life dedicated to rum, milk and coprology," as was engraved by the farm writers on the plaque.

As I understand it, this plaque was originally intended to be presented during the semi-annual farm writers' gathering at Unifarm. One night of the annual meeting is set aside in which we gather in the presidential suite, and make off with everything edible, to insult the president and generally tear things to pieces. Sometimes we even quote him correctly. I headed off to Edmonton to cover this meeting and, as happens about once every three years, my spine went out of joint and pinched a nerve. It first began to do that when I began forking cow manure for Bill McRuer. I arrived in Edmonton and sprawled out in a hotel room for a day, and finding I couldn't navigate to the meeting, grabbed a plane home. After a couple of days lying straight out on a bed, my spine snapped back into place.

There is a big bull on the Alberta Farm Writers Association "tombstone" plaque and that might qualify it for a place in the front lobby of the new *Herald* building up near the zoo. I can hardly wait to get up there because things are so crowded around the newsroom in the present *Herald* building there isn't even room for the farm editor and his eighth of a ton avoirdupois. In the fourth move in 20 years out of a comfortable suite of offices to make room for other more-exalted writers, I have been assigned a cell-block on the eighth floor.

There's another Schmidt who has turned out to be a pretty good farm writer. In fact, some say Dallas Schmidt is a better columnist than he is Minister of Agriculture. It must have been the influence of Roy Farran. He made a flying visit to Dawson Creek but was unable to stay for the three-day writers' tour. Somebody must have tipped him off about what we did to former agriculture ministers while on tour. He swapped a few yarns, had dinner and went back home and wrote a thoughtful column about farm writers. He concluded by saying: "Writing this weekly column has given me a much greater insight into what farm writers go through to produce their stories. Because I usually write an opinion piece, I don't have to dig to, get information like they do. On the other hand, the process of putting out a column is time-consuming. Most of all, I know firsthand the responsibility of having the freedom to express things as the writer sees them. It's something Alberta farm writers don't take lightly."

To which I can only add: Never try to get serious about anybody or anything or you'll never get to read your own tombstone.

Schmidt freely admits that he is not always right, but it always pleases him when people are sufficiently concerned about various issues to think about them and express their views. Over the years, not everyone agreed with Schmidt and took the time to write in to say so:

COMMODITIES IN SEVERE SLUMP
– *The Calgary Herald* – July 30, 1986

Re: columns by John Schmidt in the *Herald*, "No bleats as Ontario sells Crown Land to farmers," July 16, 1986, and "Recreationists' arithmetic further saps their case," July 17, 1986.

So now Schmidt, who has been constant in his berating of those opposed to the lease land public give away as being over-emotional simpletons, has himself succumbed to an even worse human frailty, vindictiveness. We are all nothing but a bunch of fence-cutters, cow shooters, and beer-bottle throwers according to him. One can only speculate that this sudden frenzied attack of anger by Schmidt is the result of he and his rancher friends having been told for now at least, a word they are not used to hearing, the word "no." I think that what Schmidt has to say in his column about the land-sale issue is a lot of self-serving baloney. I also think that using his position at the *Calgary Herald* as a podium from which to spew forth one-sided propaganda and insults is despicable.

In one article he talks about the Ontario government expropriating, if that's the word, at $2,500 an acre, some 23,000 acres of farmland. The government now wants to get rid of the land but there are no takers. The farmers are renting the land back at a pittance. Now Schmidt is beside himself, revelling in the fact that there are no bleats of protest from the fresh-air-fraternity.

In his second article, we are treated with lessons in arithmetic and economics Alberta-style. Economic reality is what it's all about, he states, big juicy steaks. The economic reality is that commodities of all kinds are in a severe and protracted price slump on a global scale not seen in decades. Putting more land into production, or raising more beef, is hardly a solution. What is needed is a shakeout in the food-producing industry.

I will be happy to define the difference between a rancher and a businessman for Schmidt. A businessman is someone whose business I am not forced to support through government imposed taxes and price supports. As a taxpayer, I am getting sick and tired of the politicians waving the magic goodie wand and creating millionaires out of whomever they cozy up to, while the rest of us get to pick up the tab and hang around until it's time to decide who is going to be in charge of the next selective giveaway binge.

Bernard Seefeldt
Cowley, Alberta

One of compensations of Schmidt's longevity is to see the views he advocated years before eventually being embraced by those who originally opposed them.

ONTARIO HOG BOARD REVERTS TO THE PAST
The 40-Mile Country Commentator – March 17, 1992

Canadian farmers, who now use 100 marketing boards, fear for their continued existence. Evidence of this was seen in that big rally in Ottawa designed to stiffen the federal government's resolve to continue its policy of supply management. Under pressure from the majority of GATT signatories, the federal government cannot give 100% assurance to farmers. This powerful world trade alliance contends that Canada's supply management (or orderly marketing) system has no place in the 21st century. Member countries don't very often talk back to GATT.

It is therefore ironic that, despite these government assurances, a number of marketing boards are in the process of writing their own obituaries. This has nothing to do with GATT pressure from without; it has a lot of do with pressure from within. The Ontario Pork Producers Marketing Board is one of these. It is going back to pre-Charlie McInnis days. McInnis was the architect of that marketing board. McInnis "sold" the board to producers under the rallying cry that the big meat packers were paying "under-the-table grease" to sunder collective bargaining for price.

When the marketing board took over all sales and began directing hogs to centralized yards and mixing the good with the poor, this reduced the incentive to produce quality. Packers fretted that they couldn't get a supply of hogs to start kills on Monday mornings. The marketing board took over payment to producers.

Why didn't the packers go elsewhere to buy the quality they wanted? There were two reasons. Other marketing boards forbid hogs from crossing interprovincial borders. Hogs from the United States were forbidden entry to Canada due to several diseases (cholera and pseudorabies) that Canadian hogs didn't have. Some Canadian hog industrial spies went in to the U.S. recently and found that:

1 Some northern states were on the point of becoming disease-free.
2 Some states plan to set up electronic auction systems the same as Canadian boards use.
3 Americans are now producing a leaner hog.
4 There are a number of giant contract production companies producing quality hogs.
5 Maple Leaf Foods (formerly Canada Packers) now controls more than half the Ontario slaughter capacity.

The Ontario spies came to the conclusion that if Ontario producers didn't start giving the packers what they wanted they'd find themselves on the outside looking in. In a move, which must have Charlie Mclnnis turning over in grave, the Ontario marketing board has now gone back (except in a few particulars) to pre-Mclnnis days. This has happened in the last six months I was told by Ontario marketing board people and Ontario packers attending the recent annual meeting of the Canadian Meat Council at Banff. Producers will be allowed to make contracts within and out province. Transport charges will be carried by the packers instead of the marketing board to allow direct runs from farms to plants. A special quality and weight-range program is to be put in place for hogs preferred by the packers.

From a personal point of view, this is one more about-face I have seen in my lifetime. As a wide-eyed young farm writer, I watched the Mclnnis juggernaut proceed about the country convincing farmers that crows are white. He painted the packers black. At various times when I stood aghast at this Mclnnis "miracle," many hog producers thought my criticism was traitorous; that I was a lackey for the packers. I now have the pleasure of outliving the criticism. With the help of a wise old mentor who taught me a thing or two about analysis, I was able to analyze the kind of problems the marketing board would pass on to producers' sons and grandsons. It took the board three decades to catch up with this analysis and institute the big about-face. Oink! Oink!

On those occasions when Schmidt reflects in print on the events that occurred to him over the years, periodically he provides ideas for doing things in the future. While Schmidt discusses work exchanges from a somewhat personal point of view, it may well be that the widespread application of this concept could forward a more global perspective and foster a greater understanding between people living in various countries.

EXCHANGES PROVE USEFUL
- *The Regional* - Week of May 30, 1994

One of the perks which keeps agricultural scientists on the job is assignments in foreign countries to enlarge their knowledge. Many of them have the presence of mind to absent themselves from the swift completion of their appointed rounds in Canada's fearsome winter to head south to summery weather. In some cases heading for warmer climates is essential to speed up experimental work on new plant varieties. It is possible to grow an extra crop or two in a year.

Two-way exchanges are becoming the norm for Canadians in other fields, too. A recent note from Glenn Conroy of Armidale, NSW, Australia, has got me thinking about this trend. He is manager of the Centre for Agricultural and Resource Economics, the tops in its field, attached to the University of New England there. Conroy was formerly in the promotions division of the Australian Department of Agriculture at Brisbane. Along about 1984 he got in touch with me to suggest we swap jobs for six months to a year, as he was interested in looking into agricultural Alberta from a Calgary base. I thought this was a capital idea as I had scourged many government bureaucrats for years and here was a chance to become one for a while.

To make a long story short, the exchange never came off because a Calgary publisher accused the two of us of plotting to be gold-bricking at his expense to have a nice big South Seas holiday. Conroy was persistent in his quest and he finally did arrange for a job swap with Rob Wilson, a farm writer with the Ontario Department of Agriculture at Guelph in 1987-88. Conroy arrived from the tropics to an Ontario winter, which was colder than the publisher's abuse. Wilson liked the Australian agricultural climate so much he emigrated there and went to work for the agriculture department. Upon his return, Conroy found himself unemployed for nearly two years before he landed employment in Armidale. The Agricultural Economics Centre offered a three-month exchange visit this year to an acquaintance of Conroy from the University of Guelph, Dr. Peter Stonehouse, associate professor of agricultural economics and business, Stonehouse was extremely fortunate to leave snowed-under Ontario for Australia in February.

The threat of libel sends media people digging through their tapes and files for documentation and lawyers scurrying for their bank deposit books. Writing in the public press for six decades without incurring a libel suit is an amazing feat and it is one that Schmidt continues to enjoy to this day. In an effort to help his fellow journalist avoid libel chill, Schmidt provides the following advice:

A LIFELONG AVOIDANCE OF GOING TO COURT FOR LIBEL
Scoops (Farm Writers Newsletter) - December 1998

It may be libellous for me to say that most lawyers don't know anything about libel. But as a scurrilous, bilious old curmudgeon who has made a living with a poison pen for 60 of my 75 years and have never been in court on a charge of libel, I may have some advice to hand out that will de-ice the growing libel chill in this country.

My first encounter with libel came at age 14 when my Old Man, who owned the *Ayr (Ontario) News*, allowed me to write a column, called "Strictly Incidental," in the paper. Soon after this parental mistake, he fielded a complaint from the poor, dear virtuous, frustrated ladies of the Ayr Women's Institute. He made an out-of-court settlement by giving me a good kick in the pants, followed by this sage advice: "Don't hold any of my subscribers up to ridicule in the eyes of their peers." That was lesson No. 1. Lesson, No. 2 was: "if you do criticize anyone in a column or story always give that person a leg to stand on in the last paragraph."

The next thing for the writer to remember is not to go to a lawyer if the fateful "lawyer's letter" arrives. Try to get a third party (or go yourself) to the aggrieved reader and find out why he is mad enough to sue. I once got a lawyer's letter from an Edmonton law firm on behalf of an officer in an Alberta farm organization. This letter contained 17 mistakes and typographical errors. Without consulting the editor of the *Calgary Herald*, I sent it back to the law firm with the mistakes noted in proof-reader's marks, and told the writer I didn't know what he was talking about. Then I went to a farm writer employed by the farm organization and asked him to find out what was bothering the aggrieved. The farm writer did so. He said the aggrieved would be happy with a short apology and correction, which was done, and neither the lawyers nor editors were involved.

If you are working for a newspaper don't go to the company lawyer to handle a libel suit. Call the Canadian Community Newspaper Association office in Toronto and ask for the name of the law firm retained for libel actions in your province. The CCNA has retained a firm of libel specialists on an annual basis. These lawyers are good and fairly reasonable, while most lawyers are weak on libel. I did this in my last case three years ago after suggesting to the newsletter in which an offending column appeared that we go to the aggrieved and allow him to write a rebuttal column in the newsletter.

My advice was disregarded by the newsletter publisher, but he agreed to retain the weekly newspaper lawyer. After only five minutes on the phone, this expert reached an out-of-court settlement, which was to allow the aggrieved to write a rebuttal column! There is a concluding piece of free advice. Keep your fingers crossed. A libel suit may be launched against you any time up to age 100. Also, if you have your business or house insured for public liability and you can't prevent the case from getting into court, go tell the insurance company of the suit, and ask them if they would defend you against that liability. The insurance companies have top lawyers, usually better than those of the aggrieved.

Lastly, lawyers usually send out registered letters. Refuse to pick up such a letter at the post office. The lawyer will think you are dead.

In addition to writing thousands of columns, Schmidt takes great delight in researching and writing books. He has published three and is in the process of writing another. A born writing machine, Schmidt gives some insight into the process of writing in a column he wrote about the book that he was commissioned to write for the ranchers:

W.S.G.A. HAS LOTS TO PROSELYTIZE ABOUT
Western Stock Growers Newsletter - 1994

"An Experiment That Worked," the new book commemorating the approaching Centennial (1996), is the second venture at breaking into print by the Western Stock Growers' Association. The first, of course was the work of Alex Johnston of Lethbridge, for the 75th Anniversary: a book called *"Cowboy Politics."* Johnston, who was a range management scientist at the Lethbridge Research Station, was an amateur historian who did voluminous amounts of research on places, dates and times by reading the minute books and other historical data of the association. His was a straightway writing of an orderly nature. He did not adopt the oral history technique.

However, in reviewing some of the historical aspects of the stock grower organization, I found it was a varied and variegated outfit made up of persons with a wide spectrum of interests, methods and ways of doing business. When all is said and done, though, these interests all homed in on one thing: avoiding the socialistic tendencies that have invariably invaded most of the other farm organizations. These tendencies have likewise resulted in the formation of marketing boards, which, in the final analysis, are experiments. Although some marketing boards have been in operation for long periods, they are simply nothing more nor less than experiments which don't work. They have to be changed and varied and tinkered with from time to time.

In early December, Ralph Goodale, protégé of Otto Lang, new Federal Minister of Agriculture, dropped into Geneva to a GATT meeting to defend the interests of Canadian farmers for their espousal of supply management and marketing boards and found himself next thing to a pariah. Everybody was mad at him. He came back to Canada a chastened man and ready to agree that he might need a brain transplant. Somebody send this man a copy of *"An Experiment That Worked."*

From day one the cattle fraternity decided to run its own experiment: avoiding the experiments of the socialists. I tried to get between the covers of the book the story of the people who thought it was in the best interest of the beef business to make their experiment work. Not too many people are saying they are wrong.

This book could be regarded as a vanity book for the stock grower fraternity. But I hope most don't see it that way. I hope they will see it as a "proselytizer," which they can use as a "bible" for preaching their "gospel" among the unwashed in the other farm organizations. Yes, readers, this is a hard sell. WSGA members ought to see that young cowboys come up reading this book to help give them a purpose in life, something to shoot for. They should also see that their neighbours who have put their trust in "supply management" receive copies. Those of that faith might find therein some clues about what has gone haywire with this flawed concept—a concept that recently had to be defended by the use of street demonstrations and picket lines. Peddling the gospel to the unwashed may even mean going door to door.

All during my years as a farm writer, I have had little difficulty in getting words down on paper—possibly many of them without too much thought behind them. When the George Ward-Roy Clark axis told me in December 1991 I had been dragooned into this writing chore, they gave me very little direction about what to write. This was the first "dry" period I ever experienced in my life. I sat around three months without being about to write a word. Tom Gilchrist of Milk River broke the logjam when he sent along a bundle of his papers. From then on I had fun compiling the chapters—and, further, was able to complete the writing in about a quarter of the four-year time limit at that.

One of the more interesting aspects was my discovery about commercial cattlemen and their dedication to brands. I had always associated brands and branding with a means of supplying the famed prairie oysters or a hobby like stamp or rock-collecting. But this isn't so. Among the readers I have spoken to, I have found they consider the chapter on branding and brands to be of high interest. As it was explained to me, brands denote ownership and pride in ownership. Many ranchers who go out of cattle temporarily or permanently still keep up their brand registration. A brand is more than a status symbol. It denotes a stake in a great and romantic industry.

I am told that a lady in Claresholm has an original copy of one of the 37 copies of the first brand book issued in 1888. No amount of money will persuade her to part with this historic keepsake. She keeps it in her safety deposit box.

Despite being before the public for virtually his whole life, Schmidt has a rather shy personality. He is far more comfortable on the asking rather than the answering side of an interview. Even though he attends hundreds of meetings and speaks to thousands of people, he makes every effort to avoid being a public person.

SHOE IS ON THE OTHER FOOT FOR THE SALTY SCHMIDT
The Calgary Herald by Rod Edwards (The Canadian Press) – August 19, 1980

John Schmidt was on holidays when he had the tables turned on him. After doing many interviews himself over the years, he was interviewed by a hard-nosed, no-nonsense farm writer. The interviewer was Rod Edwards of Winnipeg, who covers Western agriculture full-time for Canadian Press. Schmidt says the experience was bone-chilling, and wonders how people interviewed frequently manage to survive the impertinence. The interview is reproduced here to show how he bore up under Edwards' perspicacity.

CHANCELLOR, Alta. (CP)

John Schmidt dropped his elephantine frame onto a couch that groaned under the burden. "You wanta interview me?" mused the legendary farm reporter, clutching his favourite rum-and-cream in one hand and supporting himself on the other to redistribute his ample weight. "Whaddya wanta do that for?" he bellowed with a laugh. The shoe was on the other foot for *The Calgary Herald's* salty farm columnist, who stirs up a hornet's nest with the things he writes and does but finds his notoriety "embarrassing."

"I just like having fun and sometimes your fun leads you into some pretty stupid situations, I guess." He has dumped 254 pieces of his telephone on a government minister's desk to protest telephone solicitations, roasted politicians and bureaucrats in his column, always taking direct action when he thought it necessary.

"I know I have a pretty right-wing, hard-nosed conservative approach to life," he snorted, looking a little less imposing without one of the outrageous hats he usually wears. "But, by God, I think that's what's needed in this country!"

It's that obstreperous, defiant character that has kept Schmidt in the thick of things since he started in the business, the last 20 years in Calgary where he landed after leaving *The Kitchener-Waterloo Record* after 7 years there as a farm writer.

The 57-year-old ("I'd rather describe it as going on for 100") started out as a linotype operator at the family's weekly newspaper in Ayr and even took a swing at farming for a while. But after jumping to Calgary, he grabbed the first opportunity to write farm news for *The Herald* in 1958 and has terrorized the newspaper's deskmen ever since, fighting to get more agriculture copy into the paper.

"Anybody starting out in this business better have a thick skin and a big pair of steel-toed boots to kick the (censored) out of the desk," he says with the inevitable chuckle that takes the sting out of his outbursts.

In writing about his old friend Schmidt recently, Henry Koch, business editor of *The Record*, said he is always out to get the "bad guy." "His 'poking around' has uncovered scandals, injustices, frauds and wrongdoings including questionable spending of public funds," said Koch. "He was an investigative reporter long before the term was invented." He even had a hand in changing a federal law that said people had to pay $2 for a license to own and operate a radio.

"An inspector came into his house in 1949 and found he had no radio license," Koch explained. "John T. (the T stands for Thomas or Trouble) packaged the old radio, complete with upright speaker, the summons and $5 fine and dumped it in the office of the justice of the peace in Kitchener. He also wrote a stinging letter to the federal authorities about invasion of privacy and telling them where they could stick his radio. Shortly after that radio licenses were abolished.

"He's had some of his columns pulled by *The Herald* because of their earthy humour or indelicate wording. "I write some terrible bull at times—but I try not to offend anybody too much with it," he said, not sounding too remorseful or compliant.

Schmidt and his wife, Margaret, a teacher at Standard, have traveled extensively, taking in countries from Russia to Australia, Israel, Iran and Afghanistan, but he didn't make his first trip out of Canada until 1967. "By God, that was fatal because I've made at least one trip a year out of the country ever since. This has been the greatest thing that ever happened to me for getting perspective." Those travels provide another avenue for Schmidt's mischievous character. He takes great delight in puzzling his friends by sending little notes in envelopes postmarked in Canada but bearing the name of a hotel in some faraway country.

He escapes to his home in this village of 19, located about 80 kilometres east of Calgary where he's referred to as the Exchequer of Chancellor because they don't go in for ceremony there and don't have a Chancellor of the Exchequer. "I live out here where it's quiet. The city is getting to be just a big pain in the.... You can't move in Calgary any more." Schmidt shows a more serious side to his tough-talking, devil-may-care attitude when he reflects on what mark he hopes to leave. "All you can do is pass through the world and get a few things written down. Maybe somebody'll pick them up."

Then it was time for another rum-and-cream.

Notwithstanding Schmidt's best efforts to avoid the glare of public attention, from time to time he simply cannot avoid the accolades that appear in print. The following column, written by a fellow journalist and publisher of a weekly paper, is an example of the esteem Schmidt enjoys from his peers and readers.

BOUQUET FOR SCHMIDT
Jim Nesbitt of *The Brooks Bulletin* – January 27, 1993

In recent weeks I've made mention of two columnists in daily newspapers who were somewhat prolific in their contributions on both daily and weekly basis. Right in our own backyard, I've come to realize we have the champion of them all—none other than John Schmidt. John first burned his fingers on linotype metal years ago on the family-owned weekly in Ayr, Ontario.

When he came west to *The Calgary Herald*, he wrote daily columns for a period of 28 years, and even, if he wrote only five days a week, that's over seven thousand columns! That doesn't include the weekly column he now writes for a number of weeklies, nor the three or four books he has published... with another book in the offing. One of John Schmidt's admirers was the late Mario (Buck) Valli who provided The Bulletin with a weekly column for a few years. Buck once said he found writing a weekly column was a chore, and he simply couldn't understand how John could write one every day. I don't know either. There was a tour of the irrigation area conducted by the Eastern Irrigation District (E.I.D.) one year and John was one of the visiting newsmen. I never saw him take a note throughout the day, although he might have done it when I wasn't around, and he ran three columns as a result of that tour.

Yes, John, you can take a bow!

In addition to the public praise he often receives correspondence from strangers offering their compliments or thanks and occasionally a word of rebuke. Once in a very long while, however, a letter will arrive from a friend that is pleased with Schmidt's work. It is interesting to note that while genuine praise from friends is rare when one is received it is truly moving, even for a would-be curmudgeon. One such letter was sent in response to the following article that Schmidt wrote:

INFORMATION, PLEASE, ANTONIO LAMER!

This is a disclaimer: I am not a member of the Canadian Alliance Party—or any other political party, including the Rhineros Party.

I do, therefore, offer my services as a critic of Antonio Lamer, a retired former chief justice of Canada, who has attacked the Alliance party for "yelping" about a recent alleged odious Supreme Court of Canada decision.

Lamer offers the unsolicited advice that the elected Alliance MPs need a basic legal crash course to teach them that supreme court decisions are inviolable and are not subject to criticism or explanation. This is notice that as a taxpayer who is paying Lamer's pension, I would like him to give me a crash course. I need to know what right the Supreme Court of Canada has to reject a petition without reasons by the Alberta Barley Commission. As he is no longer chief justice, Lamer is now free from his traditional vow of silence about supreme court decisions.

The refusal to hear the barley commission's appeal affects the constitutional rights of 38,000 barley growers. In 1993, these growers voted to bring before Mr. Justice Francis Muldoon of the Federal Court of Canada an application alleging the Canadian Wheat Board Act breaches three rights granted them under the Charter of Rights and Freedoms. The three charter rights the growers allege the CWB Act denies them are: freedom of association (in section 2), mobility rights (in section 6), and the right to equal treatment under the law (under section15).

They were forced to spend nearly $2 million in legal and court costs to appeal Judge Muldoon's dismissal and the Federal Court appeal division's dismissal. When the case was appealed to the Supreme Court of Canada, the high court refused to hear the appeal. Worse still, that court relied on its long-held burlesque that it did not need to give reasons for dismissal. The decision was handed down, ironically, on the Ides of March, 2001.

The first thing I want to know is why this decision took longer to resolve than the Seven Years War in Europe? Did the file get temporarily lost in a pigeonhole in darkest Ottawa while the court dealt with minor frivolous cases that affected only 38 people rather than 38,000 barley growers trapped in the CWB designated area?

In the Seven Years War the people knew what was happening on a daily basis. In the barley growers case, those affected knew nothing until the case surfaced in a dead end (after seven years before the courts) with no explanation from the nine justices. This case had a bearing on other CWB cases before the courts. Fines as high as $7,000 we involved in some of these cases.

I want to know if it is not a coincidence that the high court decision was made at or about the same time as the publication of the book *"Jailhouse Justice: Canada's Story of Shame,"* by Don Baron of Regina. This book rounds up the stories of grain growers versus the CWB and what is happening to those who are fighting for their rights and freedoms to market grains like those outside the CWB designated area.

Lastly, I would like a crash course on whether, as Clifton Foster, General Manager of the Alberta Barley Commission of Calgary, believes it was a travesty of justice for the court refusal to let both parties to the dispute know the reason for refusal. Where is the precedent for the refusal? Both sides are taxpayers.

A democracy cannot harbour a court whose slogan can be summarized as: Like it or lump it. We say so. Therefore, it is so.

This article prompted the following:

ACCOLADE FROM A LAWYER FRIEND

Dear John,
Re: Article entitled "Information, please, Antonio Lamer!"

John, You have done an absolutely #1 job with the attached. Are you sure you never studied law?

Sincerely,
El Supremo
aka: Eugene Kush

Schmidt continues to enjoy a wide readership, writing for a number of weeklies in Alberta and Ontario. A never-ending source of pleasure for him is meeting people from all over Canada who introduce themselves as being his readers. He often gets the opportunity to smile at farm functions when speakers, like Jack Howell from the Alberta Department of Agriculture's Call of the Land program, acknowledge the far reaching influence he has that extends far beyond his column's circulation area. As a self-educated man, he also takes delight in knowing that his work is periodically the focus of discussion in classes held at the University of Calgary. Throughout his career, Schmidt has always helped to promote the work of other writers as a letter of thanks, dated July 8, 2001, clearly demonstrates.

Dear John,

How am I going to stay my sweet, modest self when I keep getting phone calls of congratulations on my education, and even more important, book orders from all over the province? This morning our minister, somewhat tongue in cheek, announced from the pulpit my fame but not my fortune yet, and I basked in whatever while he went on to tell all and sundry I'd rather by rich than famous.

Who in their right mind wouldn't? Even if it does sound a little crass. But the point is, I get calls from towns I've never heard of as well as some I have and they read about this book in John Schmidt's column and want to order one or two. How many papers carry your column and did you know you were so widely read? I don't know how many women read the column or whether they just do the phoning when their husbands tell them to, but I do want to thank you for all the advertising.

Personally, I am amazed at the success of this book...

Your friend,
Mary Jo Burles
Pincher Creek, Alta.

In line with his thinking, the last word on Schmidt belongs to the reader... so thank you Mary Jo. Mary Jo is only one of hundreds of people who Schmidt corresponds with as evidenced by the following letters and columns he has written about the virtues of keeping in touch with others and putting things down in writing to prompt changes.

CHAPTER TWO

Cross-Fertilizing Ideas

CORRESPONDENCE ESSENTIAL INGREDIENT

Letters are an essential ingredient to Schmidt's writing and integral to his life's philosophy, which is that corresponding with people is how to get things done. A fervent individualist, he dislikes mob action but never wavers from expressing his considered opinion when writing to those responsible and strongly encourages others to do likewise. He believes that writing things down not only provides the writer time for careful reflection on his or her position, but, more importantly, instils the writer with a sense of personal responsibility for what he or she has written.

Over the years, Schmidt has accumulated a remarkably thick correspondence file. But, rather than include letters from well known personages such as Queen Elizabeth, John Diefenbaker and, yes, even Tommy Douglas, the small sampling presented here are much more in keeping with Schmidt's character. Never one to pay too much homage to the high and the mighty, he is by far more interested in promoting the well being of individuals who, in his view, are being unjustly treated by others or the system in general.

Drawing on his uncanny memory (that now reaches back for a very long time) or ideas gleaned from his astonishingly large network of friends and associates, Schmidt never fails to offer a "seasoned" point of view in his columns as well as in his responses to his many correspondents. Always ready to assume responsibility for what he has written, he not only encourages his readers to write to him, he provides them with space in his column to voice their opinions publicly if they wish, whether they agree with his views or not. Rather than merely having information flow only in one direction, Schmidt relishes the opportunity of thrashing things out into the open to flush out the truth. An amazing attitude for a columnist who has laid bare his opinions for so long and one that would serve the democratic system well if practiced more commonly.

ADVICE TO KIDS – WRITE TO AUTHORITIES
The Calgary Herald – September 7, 1968

Last Saturday night I was offered an opportunity to improve my education by attending the Grade 12 graduation exercises of the high school in Duchess,

Alberta. It was an honour for a high school dropout to be asked to address the graduands. And it gave me a chance to get out to talk to people to find out what's new. In view of the fact that high school may be closed down in a program of centralization, I had a few suggestions to offer, which went as follows: Write To authorities.

Nobody should believe his or her education is ever complete. Those pieces of paper (those Grade 12 graduation certificates) are merely pieces of paper, which say you have fulfilled certain requirements that a group of persons in the department of education at Edmonton say you shall fulfill. In other circumstances that paper on which they were printed might have been used to print an underground hippie newspaper or wrap around a box of corn flakes. The thing you have to remember is never to become too smug about them. You will move on to higher schools and likely collect some more pieces of paper— but still your education is never complete.

To be the complete man or woman you have to be aware what else is going on around you. And if you don't like the things you see get up and say so— before it is too late. There's no reason we in the newspapers should be the sole public critics. You came to school to learn to read and write. Put the talents to use. Sit down and write to your MP, your MLA or even your school board member once in a while and let him know how you feel about things.

I understand in a couple of years there's a good possibility this school will be closed down. Probably the authorities putting on the pressure for this move try to let you down easily by saying "phased out," as if it were a normal thing to happen to it. The kids will be taken on a school bus to Brooks and lumped in with a lot of other faceless people, some of whom will be teachers. To me this is inherently wrong. I don't know why it's wrong.

My definition of education is "the extension of knowledge you gained at your mother's knee." Think that one over when you have a bit of spare time. If schools and teachers were to suddenly go out of existence, who would teach you? You'd be taught at home the best way your parents knew how. Our civilization thrived on that system for hundreds of years till we came to be affluent—then they went out and hired somebody to do the job. It is my opinion this was when the generation gap started to develop—and our parents have become dumb, square and stupid and everything else they shouldn't be. Actually, it was a situation of their own making. The civil servants have usurped parental responsibility—and they must blame themselves for the results.

When this school is closed down it seems to me—figuratively—the young people will be getting farther away from their mothers' knees. They'll probably acquire more learning but less education out of it all. The point is: if you don't let the Establishment know one way or another how you feel about the way things are being run, you can often be cutting your own throat. When you get "good guys" like Don Tarney, last year's principal, and Noel Crapo, this term's principal to Duchess to teach, you should make some attempt to hang onto them and the school.

It seems to me some of the education authorities have lost sight of the real purpose of education when they want to close a school like this. Apparently they want to remake the education system in their own glorified—but dull— image. They don't know it but what they are really causing to be turned out is more and better employees for the glue factory, more compulsory union members and more contributors to the Canada Pension Plan. Despite some of the handicaps young people are under, however, I don't think they will suffer too much. From any contacts I have had with kids, I find them pretty smart. Of course, there are a few too smart for their own good, but what generation doesn't have a bunch of wise guys around?

The generation before mine was smart enough to invent the atomic bomb. This new generation has to be smart enough to keep us all from getting blown up by it. Some of you are going to be like me and drop out of school now. However, that shouldn't stop you from getting an education. In fact, I advocate before you get going steady and/or married you take a year and take a look at this country to observe all the exciting developments. It will give you a vision of a country on the threshold of greatness. Then you will want to do something about making it a bit greater, you'll find you are not equipped and you'll want to go back to school or university to develop some skills.

If this were 30 years ago most of the boys and some of the girls would be going back onto farms. Only a small percentage of this graduating class will be doing that. Most of you are going from a rural area into a wider society. When you go into that wider society, don't sell your heritage short. Never be afraid to stand up and say you were born on a farm or in a small rural town and that all you know you learned here. I hope some of you—quite a few of you, in fact—will decide to go into agribusiness vocations or professions. It seems to me we are thinning out the people involved in agriculture a little too fast for the good of agriculture. We've got to have enough manpower to take care of the top seven inches of soil or it won't take care of the wider society. This, generally speaking, is what has made second-rate nations out of first-rate nations throughout history.

When you become involved with the wider society you will invariably find yourselves caught up in mob action. One of the most common forms of mob action is the strike. As you who live in the rural areas realize, farmers have been badly hurt by strikes in which they have had no involvement. Strikes have been pulled off in critical periods—periods designed to hurt innocent third parties the most. This has made for much bitterness. So when any of you get tangled up with the grain handlers' union, the railway workers or packing house workers for goodness sake don't let them "mob"-ilize you too readily into some action that is going to hurt your cousins and aunts and grandparents back in the rural areas.

The land has been good to you. It brought you this far in school. In the final analysis you will still be dependent on it when you go out into the wider society. You have an obligation to keep the wider society mindful of agriculture's important place in the scheme of things.

Schools and their financing have long been of great interest to Schmidt, who served for a year on council in his native town of Ayr, Ontario in the 1950s. Today, he is justly proud to point out that the Ayr public school building (circa 1890), where both Schmidt and his mother before him attended as youngsters, still stands. It is now many years and hundreds of students after Schmidt encouraged his community to take time to reflect on what they were then proposing to do. After listening to all the heated arguments for tearing the place down, he quietly asked if anyone had any positive things to say about the old building. Having refocused the town's thinking, more reasons were found for keeping the school open than not and the rest, as they say, is history.

Right from the get-go Schmidt railed against officialdom and seizes every opportunity to call to task any petty acts of needless rudeness. As a columnist in both dailies and weekly newspapers, he does his utmost to foster civility among his fellow Canadians.

THINGS TURN OUT BETTER THAN FARM WRITER HOPED
***The Calgary Herald* - November 21, 1959**

Dear Boss:

By the time you read this, the Royal Winter Fair will be over for another year and I'll be crossing Saskatchewan on the CPR's Dominion. Things turned out a lot better than I thought when I wrote you last weekend. I guess I was thrown off the beam a little by those Toronto newsmen at first. I was down to the horse show Thursday night and was hobnobbing with some of these people in the evening dresses and stoles and "soup and fish" and they aren't such vain and stuffy people as the sob sisters in the society pages try to make out they are. I imagine they could even go for a horse show and rodeo such as we put on in Calgary for the first time this year. Maybe not the men, but prick woman's fragile skin and you find not far underneath a sadistic lust for blood and guts— which she hopes to see spilled at a rodeo.

I was asked by a Calgary lady to write comparative views on the two shows. That would be a stiff assignment because there really is no comparison between the two shows. Calgary likes wild-west whoop-de-doo while Toronto prefers things a little more glamorous. They have the Governor-General of Canada to present a cup and the Lieutenant-Governor of Ontario to present the Queen's Guineas for the best 4-H club steer in Ontario. They eat that kind of stuff up. Torontonians were as thrilled by the RCMP's new musical ride as Calgarians were, and mighty proud of the precision developed by the 16 Mounties whose horses each made 69 jumps in performing the difficult ride through the Shanghai Cross.

Speaking of entries in the Queen's Guineas competition, that steer of Jean McInnes was not eligible for the class, as was mistakenly stated in a story in August. She emerged 6th in her class in a very tough competition. The steer was just a little over-finished.

If the Albertans found anyone around Toronto staid and stuffy it was likely the minor officialdom. Clothe a "nobody" with a little bit of authority and he seems to lose all sense of civility, spirit and wits. I have never been able to figure this out. One example, of course, was when I went in to the horse show ring without a hat the other night. I later asked the ringmaster who bawled me out why it was essential that I wear a hat. He couldn't explain, said it was in the rulebook and that's all there was to it. There must be something behind the rule. I suppose if I would have gone in without my pants he wouldn't have noticed.

The press writers were all given little green badges and told they would take us "anywhere in the show." Another farm writer and I were walking in to the horse show via the entrance where the horses entered. A cop stopped us at the entrance. Nothing would persuade him that we had a reasonable mission there. He stood there and looked blank, even to my taunt that compared to a Calgary policeman in the same position he proved to be comparatively witless. Needless to say, he was given a set of different orders by Mike Sifton, chairman of the Royal Horse Show, after a short interview by the Toronto reporter. Upon relating the experience to Harry Savage, the man who gave me the badge, he laughed and said he had gone outside the front door to deliver a package to a taxi and he had to pay 75 cents admission charge to get back in. The doorman refused to honour his press badge. It was no laughing matter at the time—the doormen suddenly smartened up after that incident.

One of my most frustrating attempts to get a story occurred one morning with a female from Ottawa who was left in charge of the 4-H club headquarters at the Royal York Hotel. I wanted to contact one of the Alberta delegation who was in a meeting. She avowed that neither the public relations man for the 4-H club council, the head of the Alberta delegation, nor the 4-H member in person could possibly see me until that night for the purpose of giving publicity to the 4-H movement. Somehow she got the idea her job was to defend her charges from any contaminating influence.

A writer from the Toronto Telegram got up at the annual meeting of the Canadian Farm Writers Federation, of which I am a member, and said that since dailies sold few papers to farmers, few of them had full-time farm writers and they didn't carry agricultural news. I contend this is a fallacy. If news coverage is based on circulation it would be reasonable to assume that since the papers had no circulation in Moscow and Washington that no news would be carried on happenings there. The impact of agriculture upon the Canadian public is just as great as the impact of Russia and the United States— and the Canadian people want to know what the farmer is doing in his own bailiwick. I cite a few of these examples to show why I think native Albertans might be given the impression that the people of the East are unfriendly, staid and stuffy. You will find the same kind in Alberta—but not as many of them.

However, nothing that this type can ever do will detract from the fact that the Royal Winter Fair is a great institution, dedicated to promoting a better Canadian agriculture. But, Boss, I wish you would send one of the women writers next year. There would be plenty for a woman to write about. She could start off by doing a story about those brassieres for dairy cows. The only thing I can tell you about them is that they sell for $21.50.

Your obedient, but enlightened, farm editor,
John T. Schmidt.

Schmidt has absolutely no immunity against humour and simply cannot resist trying to find a way to turn even the mundane into a chuckle or two. Whoever said that laughter is the best medicine must have spoken directly to Schmidt and perhaps this accounts for the long and healthy life that he enjoys.

CORRESPONDENTS' CORNER (1971)

Mr. Paul Hodgman
District Agriculturist,
Red Deer, Alberta

Dear Paul:

Thanks for sending me a copy of the April issue of *AgriScope*, your semi-occasional newsletter on farm topics and trends. You covered a lot of ground in that issue; I don't think even I could have spread as much myself!

One thing that caught my eye when I opened it up was that piece on page 10 about how I could solve my figure problem with pants. You cannot imagine my unbridled delight at being advised on how people like me with big derrieres can disguise themselves with easy-fitting, straight-legged pants that do not cup under the seat. Just as valuable was the advice that "wider legs will help balance a large derriere."

Your suggestion that a long tunic top with an interesting pattern to divert attention to the face away from this fault was something that had escaped me hitherto. Your suggestion these long tunics or vests that come below the hips to disguise protruding abdomens and large hips was also well received. Just as I was rushing off to the nearest war surplus store to buy this kind of mod garb I was disappointed in turning the page to discover the advice was written for women by Louise Sparling, the district home economist. In view of the fact men had priority on pants first could you ask Louise to send out some information about men's pants in the next *AgriScope*.

Your Sagging Friend,
John Schmidt

Despite his agility with words, numbers leave Schmidt undefrostably cold. Believing in the old adage that defining the problem is the first step toward finding a solution, Schmidt freely admits to being numerically challenged, and directs his appeal for assistance straight to the top bean counter in the country.

FISCAL AMATEUR INVITED
– *The Calgary Herald* – January 30, 1973

Mr. Maxwell Henderson
Auditor-General for Canada
Ottawa, Ontario

Dear Max:

I have recently had a great (but unsolicited) honour bestowed upon me by the 32 head of people who make up the Saskatchewan Farm Writers Association. That association, through the special pleading of C.B. Fairbairn, treasurer, at one of its annual meetings, has asked me to audit its books. I have not been able to ascertain the reason for this vote of confidence except that it is customary for organizations to have year-end audits and the Saskatchewan writers is no exception. Perhaps, living in a cradle-to-the-grave type of society as they do, they have entrusted the job to a person from a province where the citizens do get the opportunity to handle a little of their own hard-won money yet. But very little, I might add, after looking at the amount Ottawa extracts from my pay cheque. You must agree it seems rather unusual the Saskatchewan farm writers have invited a fiscal amateur like myself to take up hammer and nails to take balances off the shingles, old elevator cash tickets and restaurant napkins on which the estimable Mr. Fairbairn keeps his records at a time when you, the top professional in the land, are constantly being invited by your principals in Ottawa to jump into a pond filled with non-liquid assets.

However, before I take up my suicidal course too, I might point out a few of the idiosyncrasies of the Saskatchewan Farm Writers Association, led by Keith Dryden, and ask your advice on certain matters pertaining to their system of checks and balances. There's some urgency for having the audit done. In that organization an annual meeting may take place practically without notice, as at any time three or more farm writers get together at one location they may constitute a quorum which can vote funds to buy beer for such a meeting. Last year four annual meetings were held.

Mr. Fairbairn has promised an annual meeting some time this year at Rosetown for the benefit of Stan Sparling, who reads pastoral bulletins urging Saskatchewan farmers to grow all the wheat and ducks they can this year. For this he is eligible to pay $6 to become a member of the Saskatchewan Farm Writers Association. Because Alberta does not yet jam the broadcasts from its Socialist neighbors, his station, CKKR, is also openly heard in Alberta. Mr. Sparling was offered the chance of becoming a member of Alberta Farm Writers Association for $5 (we don't charge as much as we don't have sales tax here in Alberta), but he was persuaded to swing his allegiance to Saskatchewan when it was intimated three Saskatchewan farm writers would journey to Rosetown for the annual meeting of the Hog, Alfalfa and Chickenpluckers Society, thus becoming eligible to mount one annual meeting.

While this may indicate to you the bizarre nature of Saskatchewan farm writers' political eccentricities, my main purpose in contacting you is for an independent assessment of their suspect accounting practices. Mr. Dryden has suggested that rather than bore their auditor with statistics and burden their treasury with postage charges, they propose not to forward the usual receipts, bank statements, cancelled cheques and treasurer's report. They feel these would merely confuse the auditor. They feel it is not right to confuse an auditor with figures except to cite those of a lady member who is a perfect 33-33-33.

Ian Bickle, who works around a pool, and Gerry Wade, who is on a newspaper, insist there are no liquid assets. However, one of the group's assets is the ability to locate assets in liquid form. W.J. Bradley is expert at this. While traditional examination of the ledgers, day books, bank books and expense account vouchers has been withdrawn, the auditor has been given permission to issue any philosophical observations about farm writer accounting practices, membership qualifications and field trips farther afield than Davidson or Tisdale, the self-styled Land of Rape and Honey. D.W. Kirk has offered to write a definition of "field trip."

The auditor will naturally be required to provide the association with a list of qualifications and titles, which I am prepared to do, as there is no fee attached to the audit. As you know, farm writers are an influential group. It is this influence that probably keeps farmers broke most of the time.

My concern in writing you at this time, therefore, is to gain your professional opinion as to whether the new Saskatchewan Farm Writers Association audit procedures are likely to set a precedent. If so, do you believe they have enough influence to spread this precedent to your principals in Ottawa? I have one other request of you. In the audit of 1955, the auditor-general was able to determine that the Canadian Army had a number of horses on its payroll. Could the same procedures be used to determine if there are any horses on the Saskatchewan Farm Writers Association roster? My suspicion that there may be horses has been aroused from a perusal of the membership list. There are two people named Innes and Knowles who allege they work for the CBC. Your own experience will tell you nobody at the CBC works.

Let me have your reply soon in anticipation of that early annual meeting of the Saskatchewan writers.

Yours truly, John Schmidt
Agriculture editor and
auditor extraordinaire

Schmidt believes that physical violence is not only abhorrent but stupid, so he spends his energy spilling ink rather than blood. His personal preference for black and white to red has made him a formidable adversary. The following articles are but a few examples of those who incurred the point of his pen.

MEMO TO DELEGATES OF SASK WHEAT POOL
The Calgary Herald – January 10, 1978

You have a front man in Regina named Lorne Harasen who has a most appropriate name as he is always harassin' members of the farm press. This is odd since he is a card-carrying member of the Canadian Farm Writers Federation. However, most farm writers are able to write off his kind of badmouthing as they realize there is a shark in every pool. In his attempts to isolate farm writers from the reality of what pool delegates are doing, Harasen has blown the cover of pool management in its design to try to force cattlemen into the same mould as grain growers. Where the pool management is in trouble with Saskatchewan farmers with its marketing board stance for cattle is this:

The Saskpool has attained a near-monopoly position in grain handling in the province. It is also a political force with which to be reckoned. However, monopoly doesn't necessarily mean the pool is an efficient grain handler. It has managed to cover up any inefficiencies by using its political clout to wrest tariff increases out of the Canadian Grain Commission. The shot is paid by the farmers.

The Saskpool has achieved a vested interest in the cattle business. In a letter to the editor December 10, 1977, Harasen says Saskpool handles about 50% of the cattle marketed in that province. He evidently hasn't read a report prepared for the Mackenzie commission of enquiry into beef and veal prices by Bruce Hough and David Clarke, which reveals Saskpool yards handle 82%. There is no other place for them to go. Perhaps Harasen should dip into the Hough-Clarke pool of information before he starts running off at the mouth at other farm writers' "highly inaccurate statements."

Despite the fact its monopoly facilities handled 735,000 cattle and calves in the year ended July 3, 1977, it lost $101,000 on the operation. Now wouldn't it be fine and dandy if cattlemen were tied up like grain growers and all the Saskpool management had to do to recoup the losses would be to apply to some quasi-judicial body for a handling-charge increase? The delegate body of the pool was smart enough to realize cattlemen in Saskatchewan would give them a fight if they were boxed up like this. The delegates backed off the management's marketing-board stance when they saw deliveries to Saskpool elevators in certain areas drop off.

Cattle feeders across Canada have tended to regard the Saskpool a threat to them and the industry in general, after it was revealed the pool lost $310,000 in 1973-74 by manipulating the market by bidding against them in the marketplace. Cattlemen are also uneasy about the way the pool operates in the marketplace. The Hough-Clarke report says its terminal in Regina is the only such facility in Canada where the owner of the yard can buy and sell cattle as a commission firm. Why would an organization with 70,000 members back off from a direct confrontation with Saskatchewan cattlemen, many of whom are pool members? The answer is that 70,000 is merely a "scare" number. Alberta Wheat Pool says it has 55,000 members but only 36,000 are "active." On the same basis, a fair estimate of Saskpool active members is 45,000.

Harasen attacks Gary Jones, president of the Saskatchewan Stock Growers Association, for knowing something about the cattle business but precious little about the Saskpool modus operandi. Knowing something about the cattle business should have earned him the privilege of speaking to the delegate meeting of Saskpool as a fraternal organization delegate. After all, President Carter of the United States, who knows more about atom bombs than any other, was accorded the privilege of speaking about that subject in the Indian Parliament the other day!

I would suggest if Harasen got out and rubbed shoulders with the farmers the same way as some of the rest of his farm-writing contemporaries, he would hear their oft-voiced complaint that "the pool is so big that the farmer on the back concessions gets no say any more." I suppose it's better for him to hide in an ivory tower in Regina rather than field such complaints, as his fellows do.

There should be no mystery about the statement that the National Farmers Union (NFU) conducted a series of cattle meetings on behalf of the Saskpool. There was an interchange of speakers between the country meetings the NFU ran and the pool ran last winter. The reports and recommendations of the two organizations were practically indistinguishable. What is amazing is how the NFU financed its meetings when it had a deficit of $415,000 on last year's operations. At this distance, it's hard for me to ascertain whether the delegates dictate policy to Harasen or whether he dictates policy to the delegates. I would suggest the former applies because he makes a point of saying only the delegates may speak at delegate meetings. If this is the case, then what's he doing trying to ruin the delegates' image?

LETTER TO JIM BRADLEY
– *The Calgary Herald* – September 14, 1978

Mr. Jim Bradley,
Crusading Farm Writer,
Red Deer Advocate,
Red Deer. Alta.

Dear Jim:

A former member of the RCMP in Red Deer passed along your August 21, 1978 column in which you make a number of unresearched assumptions and ask a number of questions which merit answers.

"Mr. Schmidt is considered by many to be the best farm writer in the country," you say. In all humbleness I defer to Thaddeus McMurphyvisk of Hnough, Michigan, Jim Bradley of Red Deer, Jim Romahn of Kitchener and John P. Schmidt of Ayr, all of whom are younger and have a great deal more energy than me, not to mention being better writers. "As is the custom of farm writers. Mr. Schmidt is a conservative soul in political outlook—if not in behaviour," you aver.

Jim, you know very well I am too young to be a character. But thanks, anyway. You have at last given me my passport to the West. Everybody is conservative here. That's why the rest of Canada is now looking to Alberta for leadership.

You say I dislike Communists. Actually the reverse is true. They dislike me, as they dislike all dissidents. On December 15, 1971, a front-page article in *The Canadian Tribune* by William Beeching, secretary of the Communist Party of Saskatchewan, accused me of being a "red-baiter." *The Canadian Tribune* is issued by the Communist Party of Canada. This is the standard tactic of trying to destroy the credibility of a critic, in this case for criticizing the top leadership of the National Farmers Union (NFU). You say I dislike the NFU. Roy Atkinson, president of the NFU, charged me at its 1971 annual meeting as being a red-baiter. His speech was reported in *The Canadian Tribune*.

NFU attempted to start a campaign a few years ago to sell unpasteurized milk. I gave this campaign some space—only to reap criticism for publicizing such a hare-brained scheme as, shortly after it began, we had a tremendous flare-up of brucellosis in dairy cattle herds. Could you love an organization which advocated such a scheme?

You say I dislike wheat pools. Once again the shoe is on the other foot. They dislike me for giving space to critics in their membership who are very much concerned about the head offices losing contact with grassroots membership.

And I dislike marketing boards, you charge. Yes, well what other attitude can I take when the Alberta Hog Producers Marketing Board refuses to allow Jim Bradley and John Schmidt to attend the annual meeting but allows farm writers Ed Schultz, Garry McMillan and Bill Owen to come? My dislike may be mirroring the views of a great many producers who have belatedly discovered the much-touted cost-of-production formulae have degenerated into political prices for their commodities.

Bradley says I like cattle barons. This is a generalization. There are only two in Alberta: the Earl of Egmont at Nanton and Lord Roderic Gordon at Rimbey. I think they are fine cattlemen. And I am charged with liking Adam Smith. You have the name wrong. It's Adam Schmidt—and he remembered me in his will.

You number among my "likes" South American nations run by military dictatorships. I was in Chile in February of 1970. I liked the fact it was 30 above there while it was 30 below in Alberta. The people were friendly and their music was appealing. There was a Communist government in power then. Farmers were complaining to us that their neighbours were being machine-gunned and dispossessed of their land. Some of Salvadore Allende's people found out I was in the country and were going to come around and explain why this was necessary—but they never showed. You can't maintain food production by shooting farmers. Chile didn't.

Now that the farm editor of the Advocate has exposed my pedigree to his readers, I shall show him how I use these qualifications in dealing with contacts on the farm beat and specifically Roy Atkinson, president of the NFU, the subject of analysis in the August 8, 1978 column. Atkinson has a public speaking technique of using straw men. He has a set speech made up of all the slogans he can remember but to vary the content he opens up with knocking down a straw man or two.

He spotted me in an audience one night at the University of Calgary and used it on me. I was personally to blame for the RCMP putting him on its security blacklist, he would have his audience believe. This technique can be used effectively a time or two but it becomes boring as a constant diet. I maintained in the column Atkinson's technique has driven away all but 8,000 farmers from NFU membership rolls. Therefore I don't take seriously the allegation of Atkinson the RCMP is harassing him. It has thousands of names on file and it has his name because of several shenanigans in the nature of civil protest and disobedience that he led.

The force has my name in its files because I stated quite bluntly a couple of times it was ruining its image by hiding in the bushes with speed traps rather than flushing out of the bushes the big game like the Mafia, dope peddlers and international spies. Another way of getting one's name on RCMP files is to travel to a Communist country. Both Atkinson's and mine are on that list. The RCMP also has a list of persons from Communist countries who travel to Canada. The force's security division asks as a matter of routine that Canadians make themselves available for an interview upon return. I am not sure whether this practice continues today, but it was in force when I made a trip to Russia in 1971. I feel such surveillance is a legitimate part of national security.

While in Russia on a tour, I received first-hand experience on how the Russian RCMP, the KGB, operates. A man from Seattle who had been employed by Boeing Aircraft Company was an avid amateur photographer with several cameras. He was observed by an "Intourist" guide shooting a picture in the Moscow airport which is a banned area. Nothing was said to him, but that night the KGB broke in to his room next to ours and stole his cameras. My wife and I mentioned this to the RCMP officer who interviewed us on our return. He gave us a knowing look.

A few weeks later the RCMP operative, who debriefed me on my return from Russia, called and asked if I would like to accompany him to Lethbridge during the visit of a Russian delegation to view a meat-packing plant there. He merely asked me to keep my eyes open. Ah, I thought, here would be my one and only chance to make like Nero Wolfe, Secret Agent X-9, Dick Tracy, Peck Barnard, Perry Mason and the Pink Panther all wrapped into one. I regarded the request no differently than accepting an invitation of Canadian National to ride an engine, from Alberta Wheat Pool to inspect its terminal elevator at Vancouver, the post office to see what happens in its mail-sorting plant at Calgary, or Joe Dokes to ride his new combine. As it turned out, the exercise was a bit of a farce. I established one thing, however: The KGB man in the delegation was not the same one who stole my friend's cameras in Moscow.

Now that I have established the rate of pay involved for gumshoeing, you ask whether all *Herald* reporters supply information to the RCMP on request. I don't know. I never had occasion to ask. You ask: "Is this service to the police reserved for visiting Russian delegations or does it also apply to information reporters dig up on the NFU, Unifarm and the average farmer in the back forty?" My answer: As a Canadian citizen, I consider it a duty to the state to report any organization or person in the world who might be breaking or attempting to breach Canadian security. My allegiance is not to the Cominform. You say as a Canadian you wouldn't. Whether that confession is relevant to this discussion I'll let reader decide.

"Do the editors know?" you ask. My reply: If they didn't then, they do now. Editors change from time to time and I feel they should be treated the same as any other reader. If a great body of Herald readers (including the editors) in a groundswell of citizen indignation protests my modus operandi, I will quit without being asked. Until then the minor critics will have to hold their peace.

You conclude with the pious desire to know a lot more about the RCMP and its policy of hiring reporters as sources of information. You now know as much as I can tell you. But maybe some advice from my Old Man, who was a weekly newspaper editor, would help you gain more knowledge. Provoked one day by a lot of questions by a 14-year-old son, he demanded: "Why don't you keep your mouth shut and your eyes open and see if you can learn anything that way?"

In the meantime, continue to bore in on the great farm issues of the day to the best of your ability.

Best regards,
John Schmidt

SHOOT FIRST, ASK LATER
– *The Calgary Herald* – April 10, 1979

The following letter comes from Bill Owen, a long-time Alberta farm writer and now public affairs manager of the Alberta Hog (Pork) Producers Marketing Board, in Edmonton: I have read and enjoyed your column for many years. By no means do I always agree or support your views or interpretations. I have accepted the fact that it is your column and you do as you please short of libel or scandal. However, the February 16, 1979 column comes too close to home for me to ignore. I am writing to object strongly to your statement that, "marketing boards make and promulgate regulations without approval of the supervising marketing councils and cabinets."

You may have taken this quote from H.K. Leckie, retiring general manager of the Meat Packers Council of Canada, but in any event the statement is totally false. Our board and other boards and commissions in this province are, in fact, fully responsible to the Alberta Agricultural Products Marketing Council and the Marketing of Agricultural Products Act. Thus we are responsible to the cabinet as well. No one says we are perfect, but we are trying to do a job for Alberta hog producers, so it makes me unhappy to see a statement that is incorrect. We would be very pleased to have you visit our offices at your convenience in order that your understanding of how we operate might be clarified, Owen concludes.

In reply, I would like to say Bill Owen's invitation to visit the hog board offices is the best offer I have ever had from that organization. I have offered on several instances to be present at the hog board offices during the annual meeting but I have been told my presence would not be welcomed. Neither would the presence of any other of the 45 members of the Alberta Farm Writers Association. The only exceptions are those employees of the board who are farm writers—a very discriminatory situation. However, I suggest a convenient time for me to visit the offices would be the two days that coincide with the annual meeting at the end of this month. Having got a foot in the door, it would then be a short step into the meeting. I have a suspicion the delegates don't know their friends from the farm press are being kept out by a hard-nosed cadre in management and directorate. The marketing board sprang from the Western Hog Growers Association. I always had an excellent working relationship with that organization, but the minute the marketing board was formed the tailgate slammed down and all farm writers were unceremoniously kept out—including you, Bill, at one point. You say this column runs short of libel. Yes, in 30 years I have never successfully been sued for libel. The columns that border on scandal have dealt mainly with hog boards here and in Ontario.

Now, let us examine exactly what Leckie was talking about in the statement he made. He said a general uneasiness is developing that agricultural products marketing councils in certain provinces, whose job is to supervise hog boards and other marketing boards, are not supervising. A trend has developed for the councils to assist the marketing boards to draft instructions to the trade and they are not inclined to upset their own handiwork when criticism appears. He then went on to say in his speech in Calgary (and I shortened this down for lack of space) that marketing boards are given subordinate powers, i.e., the power to make regulations which have the effect of law without approval of the lieutenant-governor-in-council. "A number of marketing boards commonly do this," he said. He then quoted Mr. Justice J.C. McRuer, retired chief justice of Ontario, as saying this should never be done.

And how is he able to make statement? He said the meat packers council made a survey which indicated that in a number of provincial jurisdictions—B.C., Alberta, Manitoba, Ontario, New Brunswick, and P.E.I.—hog and other boards "may make and promulgate orders without the prior approval of even the supervising marketing councils and in some cases without having to give the notice of them to affected parties. You allege this is a "totally false" statement, especially in relation to the way the Alberta hog board governs its affairs. It is my understanding the Alberta board changed the "plan" approved by the Alberta cabinet when it signed domestic hog supply contracts with two packing companies last August at undetermined prices. Now, when I come to visit your offices could you provide me with the date of the order-in-council covering this regulation?

I am further appraised that two packers, Canada Packers Ltd. and Swift Canadian Company, went before the Alberta Agricultural Products Marketing Council at end of February to appeal the hog board ruling. The council held a public hearing and came to the conclusion that it didn't have the capability of ruling on the complaint. The matter has now been taken to district court following a preliminary skirmish in which a number points of law were argued.

Incidentally, William, why aren't you keeping your fellow farm writers and rank-and-file hog producers up-to-date on all these exciting developments with some well-written press statements?

Thank you, once again, for your kind offer,

John Schmidt

Undeniably his most recalcitrant alter ego, Thaddeus McMurphyvisk serves admirably in putting forward serious messages to save Schmidt from resorting to preachy homilies. Fascinated by this character, readers delight in offering commentaries about McMurphyvisk. One of the most unusual comes in the form of a poem penned by Lois Valli of Brooks, Alberta, it reads:

McMurphyvisk

McMurphyvisk – McMurphyvisk, a name of great renown,
A politician, sage, and entrepreneur, a resident of Chancellor Town.
Is he serious or joking, I ask myself, as I read the weekly news.
If I should meet him on the street, I would test him on the spot.
I'd say, "What's the price of lamb skins today in Bokhara,
Do you know or not?"

MCMURPHYVISK CAMPAIGNS TO GET ONE-MAN MACHINES
The Calgary Herald **– June 12, 1985**

Mr. Solomon Kyeremanteng
Co-ordinator of Farm Safety Programs,
Alberta Department of Agriculture,
Edmonton, Alta.

Dear Solomon:

Thaddeus McMurphyvisk showed up one day recently with a large blueprint. When I started pawing it over, he jerked it away. "I can't allow an unlettered farm writer to go prying into my new program," he said. "I'm going to use it to try to get $10,000 from Solomon Kyeremanteng to promote this idea."

"He can't pay you for a mechanical device," I replied. "His job is to preach great homilies about farm safety."

"But I'm going to contact him anyway to get his money bags loosened up for the greatest thing to ever hit the farm safety crowd," he said. "It's one-man farm machinery: one-man tractors, one-man self-propelled swathers, one-man combines, one-man motorcycles and one-man lawn mowers."

The basic invention is actually a speech. With McMurphyvisk's invention no machine will be able to move from a standing start until a sign pops up which says: "I am a one-man machine. Only one person is allowed to ride me. The driver is the man." Then a tape comes on and blares the same message to anyone who tries to climb on with the driver. The invention is a reminder to all not to allow kids to get up on machines and ride around with the driver. This is the way many youngsters get injured or killed on farms. In the last decade there have been 457 farm fatalities in Ontario. Of these 84 were kids under 15. The ratio is the same across the country.

McMurphyvisk tells me he came up with his idea the day the neighbour's kid showed up at his farm at Hnough and demanded to be taken for a ride on the garden tractor. McMurphyvisk had seen the kid's father lugging him around on the power lawn mower and was horrified. But he didn't go over and drag the kid off and punch the father in the nose. He just made sure he had plenty of gas in the station wagon for an emergency trip to the hospital. When the kid came over and sat on the seat of his garden tractor and wouldn't get off, McMurphyvisk read his speech. The kid jumped off and ran home screaming to his mother: "That rotten McMurphyvisk won't give me a ride on his garden tractor. I'm going to slash his tires." Needless to say, McMurphyvisk was a pariah around the neighbourhood for two months. However, he would rather be a pariah than a pallbearer. He became a minor hero later when he used his station wagon to run a third neighbour's kid into hospital after the kid was thrown off a tractor when he was hitching a ride with his sister and the tractor hit a gopher hole and bounced him off under the wheel.

So when McMurphyvisk comes to see you, please have $10,000 ready in bills of small denominations for him to start his campaign for one-man machinery. Have it ready for him even if it means reducing the amount of preaching pamphlets, silly bumper stickers, phony prizes for journalists who parrot safety slogans and holding seminars to hand out more literature.

Best regards,
John Schmidt.

Occasionally, officials want their version of an issue made public, Schmidt and the Herald were pleased to oblige.

CENSUS ARTICLE MISSED TARGET
– *The Calgary Herald* – February 2, 1986

Letter to the Editor:

I read with interest John Schmidt's column of December 19, 1985 entitled "Some improper queries asked during census." I can only assume that his article, which contained numerous factual errors and misconceptions, was written in jest or in haste. I would like to provide your readers with additional and accurate information.

No census question is included to determine "inconsequential incidentals." Every question included is the result of intensive consultation with a variety of data users across the country, including business groups, social service agencies, ethnic associations, agricultural associations and governments. These organizations and groups use census information for marketing, social and economic planning, forecasting, and for shaping and monitoring the impact of government policies and programs.

As to Schmidt's concerns about "census takers" being "party hacks" and "hangers-on," let me assure him that the majority of 1986 census employees will be students and youth referred through Canada Employment Centres. They will be hired only after screening and testing by Statistics Canada.

Schmidt is correct that 20% of Canadian households will be required to complete the longer census form. However, to the chagrin of pundits like Schmidt, there is no question about bathrooms in the 1986 census. Nor will we, as Schmidt suggested, be inquiring about the sex life of the nation. I would like to assure Schmidt and your readers that Statistics Canada takes great care to ensure the accuracy of census results. This includes special procedures to account for persons who are away from home on census day, and careful checking to see that each household is counted only once. The care taken to ensure accurate coverage is second only to that devoted to safeguarding the confidentiality of information supplied by each and every respondent.

D. Bruce Petrie
Assistant Chief Statistician,
Statistics Canada, Ottawa

A self-style defender of individual rights, Schmidt tries to ensure that organizations serve the farmers rather than the other way around. The following open letters demonstrate his concerns for getting what he considers to be the best opportunities for obtaining good prices for farmers for their crops.

BIG SPLASH WITH THE POOL

Mr. Doug Livingstone
President, Alberta Wheat Pool
Calgary, Alberta

Dear Sir:

This letter is a plea to you to forget your National Farmers Union leanings and come into the 21st Century on the matter of the forthcoming removal of oats pricing from the Canadian Wheat Board. Your posturing on this issue is worthy of some clients in that famous institution in Ponoka, Alberta.

As president of Prairie Pools Inc., you commissioned a survey of 600 Prairie farmers on the matter. As a result of this poll you are claiming that 63% are unhappy with the decision of Wheat Board Minister Charlie Mayer to de-regulate oats. Prairie Pools claims it handles about 63% of the grain sold for export on the Prairies. It is no coincidence that these figures are both the same. In fact, I suspect you have cooked them to enable you to convince Mayer and his cabinet colleagues that since the pools handle 63% of the grain, therefore 63% of farmers must be against the order effective August 1st.

You have as much of a chance of getting oats back into the CWB as any sane Canadian would have if he attempted to have the federal government bring back the beloved Union Jack as our national flag or any Soviet citizen would have if he asked Mikhail Gorbachev to cancel perestroika. I submit to the theory that increased consumption of white bread made from Canadian wheat made the Soviets smart enough to demand perestroika.

Before the Decima people went to work for you, you cooked the books in another way to achieve a favourable result for your own philosophy. You widely tout the fact the Alberta Wheat Pool is backed by 65,000 members. However, the annual report for the year ended July 31, 1988, reveals there are only 48,000 CWB permit books in this province—and 144,000 on the Prairies. If you are waving the Alberta membership figure under Mayer's nose, you are out 17,000. Take off another 12,000 for permit book holders who don't use the pool and that gives AWP only 29,000 actual producers, about half what you claim.

There is estimated to be 14,000 "professional" oat growers in Western Canada. This means only about 10% of the permit book holders are oat growers. In Alberta this could mean about 5,000 grow that crop and turn it over to the CWB. The poll was cooked when Decima was given a list of 65,000 in Alberta to pick from to carry out the poll. Is it rational to carry out a marketing survey that is loaded with philosophical poison like this? Why try to make me believe that a survey on oat marketing is correct when it is carried out among people who are principally wheat and barley growers, may now be out of farming, in old people's homes, feeding cattle, driving trucks or going to university? Decima claims its polls accurate within 4 percentage points 19 times in 20. Yes, that may be correct in the crowd that you have them poll but not among professional oat growers. The only place you might get agreement that you are right is in Ponoka.

Harvey Gjesdal, one of three from the 11-man producer advisory committee to the Canadian Wheat Board who backed up Mayer, claims he registered his vote on the matter by talking to oat growers across the Prairies and they told him they wanted out. Were they lying to him and telling the truth to Decima? Be rational and believe what an elected producer tells you.

The Decima pollsters said 76% of those they talked to resent not being consulted on the matter. However, was a companion question asked: "Did farmers have a chance to vote to place oats pricing in the CWB orbit in 1947." The answer to that, of course, is no. There was no vote. The feds forced monopoly pricing on the farmers.

Some rough penciling will show you the CWB at best sells about $45 million worth of oats a year for those 14,000 growers. This is just a pittance beside the $2.3 billion worth of wheat and barley the CWB sells. This means the average permit holder would receive $300 from oat sales annually. You and I know this figure has to be wrong as only 10% of the 144,000 permit book holders sell oats. So divide $45 million by 14,000 and you get about $3,000. This figure is still a pittance and that $3,000 would represent less than what four steers would bring. So your whole exercise is overkill.

The pool with its inflated membership roster is defendant in a small claims case in Taber. The plaintiff is a Vauxhall farmer whom the Alberta Wheat Pool claims as a member but he claims his membership is spurious. He wants his $5 membership refunded and his name stricken from the roll so when you go to Mr. Mayer and claim to represent 65,000 members you are not able to claim him and others like him. The plaintiff farmer holds that under the Charter of Rights and Freedoms he has the right to freedom of association and, by inference, the right to dissociate himself from any association. The AWP has notified him it plans to fight him on this contention.

To have a judicial ruling made on a constitutional right, the AWP is going to force him to pay for two lawyers: one to fight for him and one to fight against him. The AWP will be using patronage dividends withheld from him to pay its lawyer.

This is really crazy, isn't it?

Crazier still is the fact that AWP statements and press releases on the oats issue appear to have been turned out by the National Farmers Union rather than a company with a good sound business base. Remember, the NFU is an organization of radical farmers which, at one time, declared the wheat pools as Public Enemy No.1. From that point on NFU membership dried up radically. You are the only farm leader who seems to pay any attention to its extreme philosophy.

It is quite interesting to note that when the Mayer announcement came out, the AWP and other pools had their subsidiary grain trading company jump right in to the market to take on oat sales for Prairie oat growers.

Best regards,
John T. Schmidt

Livingstone's reply to Schmidt was received and published as follows:

Mr. John Clark
Publisher, Gainews
Winnipeg, Manitoba

Dear John,

I don't know where to begin. Your May 29th issue published so many inaccurate accusations and falsehoods about Alberta Pool it is a daunting task to reply to them all. I'll make the job easier by replying to the "H.M. Derails CP Rail Plan" story written by the mysterious "H.M." first and then attempt to correct some of the more scurrilous accusations in "An Open Letter to Prairie Pools Inc." written by John Schmidt.

The first article deals with discussions between Alberta Pool and CP Rail over running a unit train test on the Taber Subdivision. It blames Alberta Pool for holding up the experiment and causing the loss in potential savings of $10,000 a train to area farmers. If "H.M." had bothered to contact Alberta Pool he or she would have been told that Alberta Pool wanted the project to go ahead but couldn't live with the conditions CP Rail had applied to the experiment. CP had worded the car allocation regulations in such a manner that rail cars would have been taken from Alberta Pool and given to the opposition.

Alberta Pool asked for revisions that would have allowed it to maintain its competitive advantage, these revisions were rejected by the railway, and CP withdrew the application. Alberta Pool still wants the project to go ahead, but not at the expense of lost business. I can understand why your employer, United Grain Growers, wanted the CP proposal to proceed. It would have given UGG more business on the Taber Subdivision, but I resent your attempt to malign Alberta Pool by providing a one-sided account of our negotiations with CP Rail.

Now on to Mr. Schmidt. John states that I and/or Decima Research "cooked the books" before presenting the results of the Decima survey on oats marketing. John should know such a statement is libellous and I suspect he is gambling that our good nature will prevent us from taking him to court. The reason we selected Decima is twofold. First, it is the most highly respected polling organization in the country and second, we wanted to hire someone who was totally independent so no credible opponent could criticize the method in which the poll was conducted.

John doesn't like the survey's message so he's attempting to discredit the messenger. He doesn't have a clue how the 600 farmers interviewed by Decima were selected yet he rails against the selection process. I can assure him the selection process had nothing to do with the Pool's membership list. Decima used a selection process designed to get unbiased results. We believe the results to be accurate and we've tailored our oats marketing policy to the desires of the majority of Western Canadian farmers.

John also criticizes Alberta Pool for aggressively marketing oats after that crop was taken from the Canadian Wheat Board. What do you want, John? First you criticize us for being dogmatic, then you criticize us for being pragmatic. Of course we're going to sell oats aggressively if that's the mandate forced up us by the federal government. We may not like the new rules of the game but that won't prevent us playing the game the best way we know how.

John also infers that Simon Peter Wynker, a Vauxhall farmer who is going to court to challenge the Alberta Wheat Pool Act of Incorporation, is being forced to be a member of Alberta Pool against his will. He doesn't mention that Mr. Wynker has been paid his member reserves twice and had the opportunity to end his Alberta Pool membership when he was paid out the last time in 1987. Mr. Wynker did business with us after that and, in doing so, revived his membership. Here again it was a case of your newspaper only bothering to tell half the story.

In future feel free to report about Alberta Pool but, for the sake of fairness and accuracy, give us the opportunity to tell our side of the story before you publish anymore unsolicited open letters.

Yours truly,
Doug Livingstone
President, Alberta Wheat Pool

Schmidt, in turn, submits for publication his rebuttal to Livingstone:

Mr. John Clark
Publisher, Grainews
Winnipeg, Manitoba

RE: IN THE NAME OF GOD

Dear John,

Twice in the last two months, I have been phoned by two pollsters, the like of Decima Research, who wanted my opinions. The first question one of them asked was: "Are you a member of the media?" When I answered yes, the female said: "We don't want your opinion"—and hung up without explanation.

Then an Edmonton pollster called and wanted me to answer some questions. When I asked who had hired the company, the kid didn't know and referred me to a supervisor—a snippy female. In answer to my question, she replied: "You sound if you have been snorting coke—and we don't want to talk to a cocaine addict." Wham went the receiver.

So much for the selective way pollsters operate.

You assert you are in the grain handling business and have "opposition." Further, you think the opposition was trying to get the jump on you in the Taber subdivision. So you are a commercial enterprise. In the second half of your letter you indicate you do not like a political decision made in Ottawa by Grains Minister Charlie Mayer. You hire Decima Research to take a telephone survey of 600 people purported to be 600 farmers on a political question so you can get the jump on the opposition in a commercial enterprise.

If I am so much confused by this kind of manoeuvring that I have made what you consider libellous statements, I apologize. Do the Liberals go to the Conservatives to get opinions about Liberal policies? Does Eatons go to Petrocan to get opinions about Eatons pantyhose marketing? Does Charlie Mayer poll fruit farmers to get opinions on how to run the grain portfolio? Does Decima Research use a list of wheat growers to poll the hopes and aspirations of professional oat growers? I can't find the answer to the last question in Livingstone's letter

However, I can see one point in his vilification of John Schmidt that will make oat growers cheer that they have escaped the clutches of the Canadian Wheat Board. It is that as of August 1st those who wish to sell oats will be able to receive cash on the barrelhead for every bushel of oats they want to deliver right off the combine. If there's one thing farmers love when they go to sell their oats it is the ability to shop around for the highest price in their own community. For their sake, I hope the Alberta Wheat Pool pays them better than the opposition, whoever it is.

As for the Simon Peter Wynker case: I would only point out I was in the Taber court house reporting when Mr. Wynker brought his action against the Alberta Wheat Pool. I failed to see Mr. Livingstone or the Alberta Wheat Pool public relations writer, Doug Brunton, there. In view of this ill-starred indiscretion, I fulfilled my obligation to all participants to the best of my ability to the full extent of the law.

John T. Schmidt

P.S. I had a letter from the Embassy of the Islamic Republic of Iran in Ottawa today which addresses me "in the name of God." The embassy has offered to forward Copy No. 1 of my book, The Man from Oil City, to the new leader of Iran when it is published—and rather than take a contract out on me he will bless it in the name of God so every Moslem in Canada will buy it or have his right arm cut off.

Being a writer true to his convictions, Schmidt is not afraid to have some of his views seen as contentious and at times even confrontational. Nonetheless, the sincerity of his approach and his integrity has earned for him many friends and admirers.

FAN MAIL

Mr. John Schmidt
The Exchequer of Chancellor

Dear John:

It was great to get your letter, John, and your comments are right on as usual. I have always been a fan of yours and your letter expressed a full appreciation of the issues. Regretfully, I get very few letters from individuals who do understand because of the tactics of the daily papers which have been captured by the proliferation of so-called environmental groups. This is unfortunate because there are so many good solid environmentalists around but they are not the ones who get the press.

It has been a difficult year, John, and many times I have wished that the agricultural writers had a counterpart in forestry. The agricultural writers write the facts about what is going on in agriculture and how I wish that was happening in forestry.

I get phone calls and some letters that say rural Albertans do not understand and they are not sophisticated enough to understand. What hog wash! Rural Albertans are very knowledgeable and it burns me to have comments like that made.

There is no argument that in the past we didn't pay enough attention to protecting and caring for our environment. I'm glad we now are.

I am holding up fairly well, John, because I know what we are doing is right.

It is the encouragement of the salt of the earth, like you, that helps tremendously.

Thank you so much, John. Keep up your writing, I never miss reading what you write.

LeRoy Fjordbotten
Minister of Forestry, Lands and Wildlife
Edmonton, Alberta

The old adage that claims that it takes one to know one is certainly in Schmidt's case. Among his most ardent fans are people in the communications field. Over the years, Schmidt has rounded up an impressive stable of fellow journalists who respect his work and say so in print.

LETTER FROM ONE EDITOR TO ANOTHER
Farm Journalist Newsletter – August 11, 2000

Dear Editor,

I enjoyed the report on John Schmidt's comments to a communications class in the June, 2000 edition of the Farm Journalist Newsletter. The students were indeed privileged to hear about the experiences of this living legend of Canadian farm writing. His comments were vintage John Schmidt, but his jocular and irascible manner obscures a truly insightful and common sense approach to writing about agricultural issues. It is an approach that is sorely needed today, as we slowly succumb to the banalities of journalistic political correctness.

We at *Alberta Beef Magazine* and *Beef Illustrated* are fortunate to have John Schmidt as one of our columnists. We always look forward to his submissions, knowing that he will invariably debunk the rationale of some dubious issue or poke holes in some inflated ego or self-righteous agenda.

Oh, yes, he still hammers away at that damnable hundred-year old steam typewriter for his stories. I swear he uses it just to keep editors, who are captives of electronic devices, humble.

I feel honoured to be associated with John Schmidt, the godfather of Canadian farm writers.

Will Verboven
Editor, *Alberta Beef Magazine* and
Beef Illustrated

Sixty odd years of hammering out copy and there is no sign of him stopping any time soon, if nothing else, Schmidt qualifies for the Guinness World Book of Records for sheer longevity. However, there is more to Schmidt's work than the sheer volume of columns and books written. He explores the fascinating world we live in through an agricultural perspective and shares his finding with his readers by employing various literary devices—most importantly humour.

If the philosophers who believe that humanity is best understood through its creativity are correct, then John T. Schmidt's life is best understood through his work. He writes thousands of letters and columns about the various aspects of food production in various counties, focusing primarily on agricultural developments in his beloved Canada. Having worked in the newspaper industry all of his life, gives him a unique vantage point for observing the evolution that has occurred in the media in the last half century. And, most laudablely, he makes great efforts to disseminate his knowledge in an understandable and entertaining fashion, thereby making important and useful information accessible to everyone.

CHAPTER THREE

Bucolic Biomass Material

HUMOR HELPS DELIVER THE MESSAGE

Being intuitively interested in people, Schmidt is deeply concerned that they too are interested in what he has to share with them. One of the keys to his column's enviable longevity is his unfailing sense of humour. Resisting the impulse to write in a dry didactic or preachy style, he nonetheless delivers the punch of his message. Encouraging readers to smile and, at times, to enjoy a good laugh enables Schmidt's information to make more of an impact. In fact, the points he makes pack more punch because he entertains the readers sufficiently to read through the whole column and keeps his readers wanting more.

Emulating Schmidt's style of using humour to maintain the reader's attention may be a lesson for all writers, particularly to those who write educational material. Doubtless, different topics warrant different treatment, but it is no secret that dry, lifeless material is unappealing and generates resistance rather than encouraging the transmission of information.

Never losing sight of the importance of the individual, Schmidt reports on one of Canada's premier events for exhibiting agricultural activities to an audience of rural and urban people. By highlighting humorous elements he is able to deliver his critique of petty acts of thoughtless authoritarianism.

LOOK MOM – NO HAT
The Calgary Herald – November 19, 1959

I committed a heinous crime Wednesday night in Toronto—and 7,000 people were horrified. My crime was riding into the ring of the Royal Winter Fair's horse show without a hat. I'm not sure how seriously my unwitting act upset the equilibrium of the show officials but it drew the wrath of the ringmaster, WO 2 T.J. Connolly. I was riding between Stewart Roffey and Elmer Rye of Edmonton on a wagon pulled by Mr. Rye's four-horse hitch of Clydesdales. These two veteran horsemen had agreed to allow me to ride in what is regarded as the biggest show for heavy horses in Canada.

When the wagons stopped for inspection of the big steeds by the Judge, WO 2 Connolly stood at the wheel of the wagon and barked, "didn't you know it's against all rules and regulations to come into this ring without a hat?" I blanched but weakly mumbled an excuse that this was the first time in more than 30 years I had gone anywhere without a hat. Later I learned that WO 2 Connolly had forgiven me when he learned I hailed from Calgary. He himself had been stationed with the Lord Strathcona Horse for a number of years at Currie.

I have heard many people speculate about the fate of heavy horses in Canada. After riding behind a quartet of prancing horses each weighing over a ton I can say this: Showing of these horses will never cease as long as men enjoy the thrill of driving show horses. Elmer Rye operates a section of land north of Edmonton. He and his father, Lawrence, who has been a Clydesdale fancier for nearly half a century, keep a few Clyde horses around the farm to do chores. The horses in the quartet are worth $600 each. They all are about 18 hands high. They stood quietly while Elmer and his assistants put on the brass mounted harness. You won't find harness of this quality in the ring today. It was made 35 years ago in England and it was supple and strong as the year it was made. "It took six weeks to break these horses into a four-horse hitch," said Mr. Rye. The secret of making them perform is to train one to follow behind the other and keep tugs tight. Kindness is the watchword of horse training; a rough driver has difficulty controlling his horses. The big green wagon the horses pulled was willed to Lawrence Rye by the head of Petrie Manufacturing Company of Toronto. It is used by all the Alberta exhibitors of heavy horses and is stored at the Royal from year to year.

Fifteen minutes before ring time all was in readiness. Up on the high seat I picked up the four lines—one to each horse—I could feel the tension transmitted from their Liverpool bits. I handed the lines over to Elmer Rye and with a quiet word he had the quartet under way. The wait in the assembly area seemed interminable. I knew both he and Stewart Roffey were tense and I, too, could feel a knot in my stomach. They cracked a few jokes with people standing around to break the tenseness.

A light horse class came out of the ring and we began to move. There were six entries—making a total of 24 horses in the ring. The Rye entry pulled into the right fourth to blare of military band on a high platform. All six hitches were plunging. I wondered how they could be kept under control but that seemed no problem for Elmer. They went exactly where he guided them. The people in the stands seemed a blur. It was impossible to spot familiar faces. Dress ranged from overalls to fish and soup.

The horses were drawn up in the ring centre and the judge quickly made his decision. It would have been a nice class to win but the judge motioned the Rye wagon to last place. The losers trotted out to the tune of a Scottish air. The winner was wildly cheered. But that's all right—there'll be another time. If the Albertans felt any chagrin they covered it up by quickly unhitching the horses and sending the horses back to their stalls.

The Alberta branch of the Canadian Restaurant Association was not at all pleased when Schmidt picked a bone with some its members in his column. Today he raises his glass to toast the splendid improvements restaurateurs in the rural areas have made. The doyen of them all was Mrs. Jean Hoare, who had a dining room in her home at Driftwillow Ranch near Stavely, Alberta and who became renowned for her grilling twenty-four ounce steaks cut in two. She later served them up in an old Air Force building, called the Flying N.

NO THANKS FOR THE MEMORY
– *The Calgary Herald* – August 18, 1962

A stopover of a couple of hours in Carmangay, during a long hard trip one night recently, has been a dissuasive force in not seeking an invitation from the Canadian Restaurant Association to tell the members why I have decided to carry my own grub henceforth when travelling. Because of the stopover at the hostel of mine host, Alex Yanosik, in Carmangay, I have also decided not to repudiate my uncle and fellow columnist, Ken Liddell. Somehow a favourable report by Uncle Ken always puts a kiss of death on an eating place as far as I am concerned. The only exception is Driftwillow Ranch near Stavely.

Maybe I am just peculiar in my eating habits. I'll gladly brave the competition of flies in a restaurant (although modern methods leave little excuse for their presence) or revolting uncleanliness in a wash room, but inflict on me a waitress who is sloppy, sullen, impolite, tardy and inefficient (yes, I have seen all these qualities wrapped up in one woman in some restaurants) and even the best food is spoiled.

Alberta's highway system is becoming one of the best on the continent with its 65 m.p.h. speed limits on good pavement. It has not been cluttered up with garish bill-boards. Campsites for respite from weary miles are thoughtfully provided. However, the good roads lead the unwary traveller and tourist into some of the dreariest, ptomaine palaces on the continent. In too many cases waitresses seem to be recruited from schools for the unwashed. Possibly we have not become civilized to recognize that serving food is an honorable trade, or that there is a large pool of talent in men waiters.

In what back alleys some of the cooks are recruited is a mystery. I know what would happen to them if they tried some of their atrocities in construction camps or cow camps. The amazing part is that a number of these same people will pay good Canadian dollars for the kind of food in a restaurant for which they would string up a camp cook without a qualm. You know the kind I mean: cold white lumps that are alleged to be potatoes, gravy that ostensibly originates in the same cans as crankcase oil, soggy and tasteless messes of green and red objects a week old that are listed on fly-specked menus as "vegetables," not to mention the brutal methods with which steaks, chops and chicken are treated.

Getting down to specific cases:

Calgary — Sent potatoes back in an 8th Avenue place as they were stone cold; received equally as cold ones from the same pot. Cook highly incensed at my recklessness.

High River — Waitress, who finally appeared, regarded me with suspicion. Stood with mouth open. Never said a word.

Taber — At breakfast, waited 15 minutes for harried single waitress to finish serving 15 people while two other waitresses sat on fannies reading papers. Finally walked out and went to another place that wasn't understaffed but they said they were just out of cornflakes. Never stocked any in the first place.

Vulcan — Cook-waitress said she was out of dinner soup; she looked as if she had been too lazy to make any. Substituted with a glass of tomato juice the size of a salt shaker; didn't know they made glasses that small before. Made a stab at cooking two

	frozen pork chops. Couldn't decide whether to half-sole my shoes or eat them.
Fort Macleod —	Got mouth set for spaghetti and meatballs advertised as menu special. Sent our "compliments" to the cook when he sent word he was "just out." Had to fight way out of the joint when cook complained to owner.
Veteran —	Hungry as a dog at 4 p.m. and was turned down for a pair of pork chops as the cook not around. It's a poor outfit that will turn down a starving man.
Bassano —	Had to shout order across the room to tired waitress, too tired to write down order, and who was commiserating with teenagers and an equally-tired cook who came out and flopped down at a table.
Strathmore —	If you can imagine it, even the peas were tough at one of the highway eating places.

Many a housewife has learned to her sorrow that one bad apple has spoiled the basket. In the case of eating places, it's just the opposite. One good place will sweeten the sour taste of many bad ones. And that brings me back to that stopover at Alex Yanosik's at Carmangay. After wetting down the dust of the trip we wandered into the dining room, from which an appetizing cooking odour was being wafted. The waiter was a cheerful young man, his son, Billy. He was bustling about with cheerfulness and despatch. He was most solicitous about bringing more coffee and butter. The meal was excellently cooked; soup hot; potatoes warm and large portions of main course; price extremely reasonable. It's the only dining room I've been in for a long time where they came around and asked: "Have you had enough to eat?" They not only enjoyed cooking but they served the food in a pleasant and distinctive atmosphere.

Confident of Schmidt's keen sense of responsibility, his readers periodically send letters to him in the hope of making public their grievances in order to rectify problems:

KISSING OWLS, LEANING COWS
The Calgary Herald – January 17, 1963

My bewhiskered friend, Bob (Fearless) Faris of Bow Island, sent off a missile the other day to say he wants to be the first in line to sign up for electricity without wires (Agricultural Alberta January 8, 1963). Bob finds the frustrations of farming almost overwhelming at times. The remedy is to sound off and, as the readers know, the last time he teed off on the banks in this column, repercussions were heard from as far as Montreal. I don't know if the banks let loose that tight grip on their vaults to the extent Mr. Faris suggested, but I have seen my banking friends plying trade more assiduously among the farmers of late.

At any rate, Bob's target is now power companies. Here is his transmission:

"The case for public power is a cause I have not been a strong believer in but now I have 'had it.' Listen to my tale of whoa and, partner, I mean whoa.

Calgary Power can think of more excuses and alibis than any management I have ever encountered. They've sent out propaganda on radio, television and newspapers and, in general, apparently have fooled themselves into thinking they are a real power company.

Let's get down to cases and take a look at some of the alibis they offer:

1. THE CASE OF THE KISSING OWLS. Power off six hours or better. I was told that one owl must have sat on one side of the transformer and one on the other side.
   They kissed and fried
   Calgary Power say they haven't lied
   But our little motors and lights
   Shore as heck died.
2. THE CASE OF THE LEANING COWS. We have been told by our power company (Calgary, that is) to fence all brace wires because our cows were leaning on the wire as they scratched their mangy hides and, you've guessed it, you smart people, another power outage.
3. THE CASE OF THE TIRED BLACKBIRDS. Four and 20 blackbirds lit upon a line, tired blackbirds, that is, and again power off.
4. THE CASE OF THE RARE AND MYSTERIOUS RAIN. We have had a drought down here in southern Alberta that has broken some mighty good records of the past so you know we've been dry, BUT we have had some little insignificant showers. Ten of them won't raise a five-bushel crop but yep, ye'r right, they stopped the power every time.
5. THE CASE OF THE DUSTY TRANSFORMERS. I also have been told that dust collects on these transformers, rain runs down, wets the dust and 'blooie,' no lights, no power.
6. THE CASE OF THE ONE-WIRE, ONE-HORSE TRANSMISSION LINE. We have a one-line transmission line, which in the case of a Kissing Owl, Leaning Cow or Tired Blackbird, stops all power on that line. Now 40 or 50 farmers have their power cut off and stays off till the owls quit kissing or cows stop leaning or blackbirds get rested. However, a two-line power line, I am told, would eliminate most of these troubles. You would think that with the price we've paid for installation and kilowatt rates we would have a two-line power line. Oh no! We just have a one-line, one-horse, one-kilowatt, 18th century power line.

I could go on for hours with these little cases but farmers know as many or more than I do. Let me sum up my friendly happy little letter to Calgary Power and say that: Never in my experience have people been served worse by a private power company than by Calgary Power. I don't mean the two men who work for Calgary Power and live here and who are run to death night and day trying to keep up with irate farmers, TV-less viewers, kissing owls, leaning cows, birds, rain, and anything that walks, breathes or flies that Calgary Power can blame for a one-horse antiquated broken down collection of posts and wire they call a power distribution system.

Got a modern dairy? Lights, milkers, coolers? How many times in the last year have you juiced your bossies by hand by the light of a kerosene lamp? Sheepman, lambing with good lights, needing lots of hot water for chilled lambs: had your power off lately? Relax after a hard day in the field or on the ranch; got a special program, hockey, football or mystery on TV or radio? Stumble into bed in the dark? And lie there cursing a sleepless two or three hours you were going to watch, be diverted and amused by the living screen. Get a modern furnace, throw away your old coal stove and have nice even well spread heat all over your house. Oops! Power off! Where did we throw that old coal stove? Everybody on the farm and ranch down this away has a good supply of candles, gas lights, old coal stoves and gasoline cook stoves. Brother, we've been educated. Almost every time our power goes off our stock don't drink all night. Why? Well, it freezes up and then the water can't run when the power comes on. Both our pressure systems have been frozen up so many times they look like a welder's nightmare.

Yes, we have AC power from Calgary Power. Now we know what AC means: it's 'alternating current,' one hour off and one hour on. Of course, we really are suckers; we have an electric water system, barn and house electric water heaters, electric cook stove, lots of yard lights, electric block heaters on tractors and vehicles, trickle chargers and when they run we like 'em. Ever since they charged us $1,500 to put the power in and charged us $30 a month to run it we are very happy. Now I hope some fool won't start at plebiscite to nationalize Calgary Power. I think that in a hundred years they'll do well and we'll have a company which will deliver power continuously."

There is no doubting Schmidt's enjoyment of food, and cheese is high on the list of his favourite things to eat. Unlike other writers who restrict themselves to nothing but dry fact, he always tries to spice up his columns with a little zesty giggle.

CURED UNDER DIRT COVER
– *The Calgary Herald* – October 5, 1963

There's one thing pleasurable about eating cheese: Every piece is, or can be, different. Professor D.M. Irvine, head of the department of dairy science at the Ontario Agricultural College, Guelph, tells me his records show there are 500 kinds of cheese. I understand there is a feast of 70—maybe more—different kinds sold in Calgary stores— enough for one or two different kinds for every week in the year. It is an amazing fact of food technology that a pail of cow's milk or goat's milk can be converted into so many different forms of food, all with a different flavour, colour, odour, shape and price. The reason this is possible is in the way it's cured. Heat it up and you get hard cheese. Slow the curing process in a cool place and you get soft cheese. Cheese ripens faster under warm conditions. The early cheese makers would seldom tell their trade secrets. They were no fools. Had the consumers learned of some of their methods they would have thrown the product to the hogs.

However, with today's appetite for more and more exotic foods, I don't suppose revealing some of these trade secrets will reduce consumer demand—at least it shouldn't in view of some of the delicious but mysterious items people pick up in the gourmet departments of stores. Here are some of the unusual curing methods Irvine outlined:

Bellelay: It's wrapped in bark for curing.

Cheshire: Is placed in a cheese oven at 75 to 80 degrees for curing. It's an English gourmet cheese.

Creuse: A product of France. It is placed in tightly closed containers lined with straw for ripening.

Forez: This cheese is made in central France. It's placed on the floor

	of a cellar, covered with dirt, over which water is permitted to trickle. Fromage Fort: Another French make, is buried in dry ashes to remove as much whey as possible.
Fromage de foin:	It is buried in hay where it remains for six weeks to three months. Oh, yes, it's made in France.
Olivet:	Country of origin is France. It is cured on straw shelves.
Slipcote:	An English make. This cheese is ripened between leaves of cabbage.

The readers will recall that in the March 20, 1963 column, I got my cures mixed up and suggested that a certain type of cheese was cured in a manure pile. This inflamed the nationals of a well-known European country. Upon discovering this cheese was not, indeed, cured in the barnyard pile, I have not eaten it since. Upon making some further inquiries concerning this rite of cheese making, I learned from a source in the dairy department of the University of Alberta that he recalls seeing nomadic tribe in one of the Balkan countries using this method, as it was the only constant source of heat they had for making hard cheese.

Some of the best cider I ever tasted came from a keg that had been stored in a manure pile for just the right length of time. That's an old Ontario custom. Actually what I've been meaning to say all along is that this is Cheese Festival Month, a promotion by the Dairy Farmers of Canada in co-operation with the National Dairy Council of Canada. They aren't pushing the promotion as hard this year because their most famous cheese, cheddar, is not in abundant supply.

In recognition of life's brutishness and brevity, Schmidt's philosophy toward writing is to inject humour whenever possible to help ease the stress of daily living. Periodically, his column offers entertainment suggestions to relieve the harshness of reality or, conversely, of its tedium.

DONKEYS, GOATS, GOPHERS AND BUFFALO JUMPS
The Calgary Herald – June 6, 1964

While many folks were drawn to the sale of bucking horses and gentle saddle stock held at Buddy Heaton's ranch south of Midnapore from many miles away, most of them showed up from neighbouring districts like De Winton, Okotoks and Turner Valley. Quite a number brought their lunches and their good-looking young folk who, on this particular afternoon, had the chance of entering in on an additional pastime—that of snaring the new crop of curious young gophers.

It may be that in Oklahoma, where Buddy Heaton hails from, gophers don't exist. With all the various animals in the Heaton rodeo stock paddock, it's easily seen this man is a fancier of pet stock. Around the place one can find a Texas longhorn steer, saddle-broke buffalo that are well known to rodeo fans, a bunch of donkeys, a herd of goats and numerous horses. What I can't understand, writes my guest columnist George Rempel of Brooks, is with all this stock to feed how he can afford to have all those gophers around, too? Thanks, George, for a picturesque description of a picturesque event put on by one of the district's more picturesque cowboys.

One of the most amazing performances I ever saw anywhere—and one not likely to ever be repeated by any man again—was performed by Heaton last November at the Canadian Championship Rodeo at Maple Leaf Gardens in Toronto. He came out riding his buffalo, Grunter. He was followed by a brand new half-ton pick-up truck with stock racks, donated by a Toronto car dealer as an advertising gimmick. The truck pulled mid-arena and stopped. In the meantime Heaton was riding Grunter full tilt around the arena, whooping and hollering. Then he made an amazing dash right into the back end of the truck. The momentum was such that Heaton kept right on going over the buffalo's head. He sailed over the cab of the truck and landed "boo-o-om" in the middle of the shiny new blue hood. Well, sir, I've never seen a nicer or bigger or well-rounded dent in the hood of any new truck like that one Buddy Heaton made. That performance alone was worth the price of admission.

The variety of styles Schmidt employs in writing his columns reflect the eclectic nature of his interests and is indicative of his relentless efforts for developing fresh approaches.

ANTHOLOGY OF POETRY
- *The Calgary Herald* - January 27, 1965

Hanna poet Ferg James has collected more than three-dozen of his best poems under the title, *Tales of the Pioneer Days* and published in book form. The book sells for $2.50 and may be obtained from the author whose address is merely Hanna, Alberta. James was born near Lindsay, Ontario, on March 31, 1890. He moved with his parents to Killarney, Manitoba at age 8. "I was raised on a bush farm and attended the old Huntley School 16 miles east of Killarney," he said. "Then I put in three years in high school in Killarney. I attended normal school and following that training taught in Manitoba three years." Schoolteachers went to work much younger then and with less training than is necessary today. Of course, this shows up in the wages. Before the First World War a teacher was lucky to get $300 or $400 a year and, sometimes, was luckier still if he collected.

Evidently James did collect his money and headed west with enough to file on a homestead in the Bull Pound area south of Hanna in the fall of 1909. Here he homesteaded, taught school and was a grain buyer for Home Grain Company and owned a dairy farm. "In 1944, after losing my right hand, I purchased a small ranch north of Richdale and ran a modest but very satisfactory spread until I retired in 1962," he said. "I now reside in Hanna, where I still have a few of the old original homesteaders to shoot the breeze with. What gives me a great deal of pleasure is when the sons and daughters, grandsons and granddaughters of these pioneers meet me in the street and call me 'Ferg'."

James had some spare time outside his many other jobs to sit down and jot down a poem here and there. He began this hobby early on. Some of the rhymes in *Tales of the Pioneer Days* go back to 1920. Most of his early ditties were written about his surroundings on the Bull Pound farm. There are verses about the Peavine Express (the mixed train which made three round trips a week between Hanna and Wardlow), his neighbours and their quirks, together with descriptions of rides over the prairies and what it was like down on the ranch. Some of the titles are intriguing: There was *The Calamity of Lost Panties,*

The Saga of the Cross-Eyed Rooster, The Saga of The Bull's Lament (this poem was used several years ago in this column), *How They Used to Bootleg Wheat* and *A Case of Appendicitis.* There are many more such titles.

Ferg James was never too busy to engage in politics. And he was never too busy to sit down and dash off a piece of doggerel about politics and politicians. He began to needle the Socreds prior to the 1935 election when they first came to power. *When Election Rolls Around* was his contribution at that time and is included in the book. He has kept it up ever since, although he hasn't made any of them as furious as *The Contributors to The Edge.*

With the Western Stock Growers Association annual meeting coming up next week in Medicine Hat, I think it would be appropriate to reproduce some versifying Ferg James did December 3, 1953:

When Stockmen Congregate

The stockmen held a meeting
To discuss the pros and cons
Of all the bovine species
From buffaloes to fawns.
They dealt with brucellosis
And blackleg met its fate
Most anything can happen
When cowmen congregate.

They came with spurs a-jingling
And big ten-gallon hats
From the hills and plains and valleys
From river breaks and flats.
They felt all but forgotten
By the boys who legislate
And the air got rather humid
When the cowmen congregate.

With freight rates soaring upward
And machinery on the hike
One industry gets settled and
Up comes another strike.
With all this vicious circle
They wonder at their fate
You hear some rangeland parlance
When the cowmen congregate.

In the halls of legislation,
Where the laws of men are made
We need a few old cowhands
Courageous, unafraid,
Who rear up on their haunches
In caucus or debate
And tell the world their troubles
When MPs congregate.

The last opponents to daylight saving time were the Saskatchewan dairy farmers and Hutterites, but they too eventually capitulated. More recently, however, Schmidt is heartened to hear some of the urbanites (the chief advocates of DST) pining about those who tinker with clocks.

DST MAKES YOU STUPID
– *The Calgary Herald* – April 19, 1967

In the provincial election May 23, 1967, I intend to vote against two things:
1. The Social Credit government.
2. DST (Daylight Saving Time).

The reason is I demand consistency in government—and the 58 Socreds aren't giving it to us. It was surprising how quickly the Manning government in a panic enacted legislation to prevent people from meddling with their minds and bodies with LSD. The government believes it is wrong for people to "turn on" by using the stuff. It is not only surprising but, inconsistent that the provincial government at the same time caved in to pressure from the time-piece tinkerers to amend legislation which will allow people to do about the same thing with DST. By using DST the meddlers want everybody in the whole province to "turn on" one hour earlier. Therefore, I am going to vote against DST simply because I don't like it. I think it's wrong for anyone to force me to "turn on" an hour earlier each day. I hope all those who think the same way will smash this daylight saving myth once and for all.

It has been alleged by certain of my friends, who know I hate Daylight Saving Time, that I hate it because I am farm-oriented, that I wish to curry favour with the cows and chickens and grain-cutting schedules. I emphatically deny this. I hate it because my experience with it for the first 34 years of my life was that for two weeks in fall and two weeks in spring I was stupider than usual and I had to deal with a lot of other people gone stupid for the same length of time. What happens when the time clock is set ahead is that the metabolic clock of every person in the province is upset. That is dangerous. It's the same as throwing everybody into a completely new environment and expecting each to adjust completely in one day. It just can't be done. Ever notice that you become miserable for days after a trans-continental or trans-oceanic jet trip? Scientists have found the reason is that the human "clock" slips a couple of gears. It throws the metabolic clock out.

DST does the same thing to a person. It sets you on edge and makes you cranky, short-tempered and stupid. It's about the same stress as subjecting oneself to having a pail of water dumped on you every morning. The scientists, who have investigated tinkering with human beings, have found the average person can readjust to sleeping and eating schedules in about two days. But they have also discovered that body rhythms such as temperature, blood circulation and the nervous system require a week or more to get back in line again. When the lower orders of mammals such as cattle are subjected to the same treatment, they react with a drop in production. There is no reason to believe that man doesn't react the same way.

In situations where people are subjected to undue stress for long periods of time, their reward is relief from the stress. The people who advocate DST don't even recognize the fact this is necessary. There is unrelenting pressure to force people into the mould on April 30 and keep them there till the end of October.

If they would do some progressive thinking it wouldn't be so bad. For instance, why not work an extra hour every day and have a three-day weekend every two weeks? In other words, institute the reward principle for subjecting the body to the stresses and strains of changing the clock twice a year.

People who wish to meddle with the clock infest even the remotest parts of the country. Several years ago some of them got loose into the Yukon. They talked the territorial government into inaugurating Daylight Saving Time—even though there are 21 hours of daylight there in summer. Imagine a government acting upon such a hallucinatory demand! They must have all been on LSD.

Attempts to provide practical solutions to the ancient philosophical problem of the one and the many inspire all sorts of wonderfully whimsical suggestions for social planning. Schmidt enjoys consulting his learned friends to find tongue-in-cheek advice on just how the world ought to turn.

PEOPLE-CULLING POLICY
– *The Calgary Herald* – August 23, 1969

The new lodge Doc Buller once planned has given me an idea for organizing a new beef breeding society. Doc, the erstwhile editor of Buller's Bull-O-Graph and Tourist's Fireside Home Companion, sat and talked for hours about the lodge. It was to be a secret society to which everyone could belong. He envisioned 1,000,000 members. "My new lodge will automatically wipe all others out of existence," he said, "as it will give the common people a chance for widespread participation and will overcome the objections of people to existing lodges." He explained that Canada has no kings, princes, dukes or barons. However, there are plenty of grand high commanders, loyal inner guards, chief psyfodlicators, outer impotentates, grand viziers, exalted knights and supreme chancellors. This situation exists through the desire of a few people to be something without doing anything.

"People are essentially joiners," he mused. "Even though they can't hold these offices they still want to socialize with their fellows in the Black Hand, the Pink Hand, the Ontario Hog Producers' Association, the Dirty Hand and many other secret societies." The lodge Doc Buller envisioned was one in which he would hold all the offices. "I will hold all the offices," he said. "If you join you will not be pestered to pay dues by the financial secretary. We are not going to have dues. There will be no regular meetings and at the emergent meetings Doc Buller's Favorite Tonic will be served with or without gin and every member in good standing may consume all of the tonic he wants. People will not be able to object to the name of the lodge because it will have no name, neither has it aims or objects. A small initiation fee of only one-tenth of a megabuck will give you a lifetime membership and all you have to do is send me the money."

My new beef breeding society could very well be based on the concept of Doc Buller's lodge. There will be 1,000,000 beef cattle enrolled. It will wipe out all the others as it is dedicated to growing beef, not pedigrees. I don't care what angle the beef grows on the cattle—as long as it grows. The only misdemeanour, which will result in blackballing, will be cattle that shrink. After a six-week initiation on Doc Buller's Favourite Tonic factory stillage, the cattle will be switched to a ration of steam-rolled barley. They must achieve gains on this ration according to a secret herd weight index, which will be kept in a locked box guarded by two Black Hands.

If registration papers arrive with the cattle they will be very carefully collected and bundled up for the next Boy Scout paper drive. However, if they shrink on the ration of Doc Buller's Favourite Tonic and steam-rolled barley rather than achieve the gains set out in the secret weight herd index, they will be given a few skin and colour grafts to restore their conformation and sent back to regular breed organizations. Every colour of cow will be admitted to the new society. There will be no blackballing of black cattle with white trim nor white cattle with black trim. Neither will brindle cows with scurs be banned.

In order to maintain high standards I will hold all the offices. I have discovered there are plenty of breed organizations where the offices are passed around among the provincial delegates so that each delegate gets a whack at the trips abroad expense-free. I have discovered many breed organizations operate on the Peter Principle: that is, the officers rise to the level of incompetence. The incompetents won't allow anyone more competent than themselves to serve nor will they vote out an incompetent as they might have to answer for voting him in in the first place. And then there are dozens of breed organizations where the officers and members sit around arguing about the constitution when they should be drinking—what else?—Doc Buller's Favourite Tonic. There will be none of that in my new beef breeding organization. I will take all the trips and if I become incompetent I'll go broke. Since there will be no meetings, there will be no wrangles about the constitution.

The real benefits of the new society will show up in the appointment of the culling committee. Since I will hold all the offices, the culling committee will be me. Culling will automatically start with people, not cattle. There would be no necessity for my new society had the present breeding organizations instituted a people-culling policy. On the surface, the only difficult program in the new society will revolve around whether to cull those cattle which don't reach the herd weight index performance before their numbers reach 1,000,000 or after their numbers reach 1,000,000. However, this problem is easily solved once it is remembered the registration papers have gone out in the Boy Scout paper collection. There will be nothing to prevent selling the culls back to the traditional breed organization where you might get some static over losing the registration paper but nobody will care much about the performance.

Since there will be no culling committee and no pedigree papers, there will be no reason for holding auction sales. With 1,000,000 beef cattle, it will be necessary to handle all the information on them by computer. Enough information on each animal will be fed into the computer to allow the computer to establish a price formula for each. When somebody with a wad of speculative money decides to go into the cattle business, he will find the only avenue is through the beef-producing business. He will send me his money and I will send him a bunch of numbers and he will be in business. Then I will drink a toast to him with Doc Buller's Favourite Tonic to Doc Buller and all the animal geneticists at Lacombe, Lethbridge and Edmonton who gave me the idea for the new beef breeding society.

The fascinating complexity of today's sophisticated lifestyle increasingly provides more options and opportunities for doing a multitude of activities. Prioritizing has become a hotly sought after skill and Schmidt generously offers a brilliant template for the readers' edification and enjoyment.

WOMEN, GOATS NEED STUDY
– *The Calgary Herald* – April 21, 1971

From time to time people ask me if I do much travelling. My usual answer is: "not enough." Agriculture is dynamic and there really isn't enough time to get around and look at everything there is to see. However, every once in a while I make up a list of priorities and try to get around for a look-see. The list is run on the same basis as the parliamentary order paper. If the boss filibusters out some of the suggested trips, they go to the bottom of the list until a later date.

Last week the priorities shaped up like this:

1. *"See if it's true that when they cross a Limousin and a Red Angus, they get something that looks like a purebred Jersey."*
2. *"Investigate malting barley prices."* Have the maltsters been able to restore the premium on barley to farmers since the drinking age was lowered to 18? This may entail a trip to the Oyen Hotel.
3. *"Make a tour to Buck Valli's place at Brooks for another story on increased money-making possibilities in sheep."* A Hereford breeder told me the other day there must be money in sheep as none has ever been taken out.
4. *"Do some more background on the T and T report (Tradition and Transition)."* That's the report prepared for the Alberta Department of Agriculture by Farm and Ranch Management Consultants Ltd. of Calgary. Interview the farmer who said he had tried to persuade his district agriculturist not to quit his job after13 years because in a few more years he would likely have enough farming experience to help farmers. The report might also give a more concise answer to the conundrum: Which is more valuable, a $14,000-a-year man driving around weighing calves for the record of performance program or a $12,000 man travelling the province probing boars? The Ribstone Rustler failed in his probe of this question!
5. *"Trip to the Souris River, Saskatchewan, bull station."* This is in response to a finance-writing colleague who wants some manure to grow a lawn. The Prairie Farm Rehabilitation Administration has 125 of its 2,200 community-pasture bulls wintering there. This should mean a copious supply of manure available but investigation of viscosity will have to be carried out first-hand.
6. *"Cross-Canada tour re: goats on record of performance."* There are only three of them on ROP and they are all owned by women. I'd like to find out if they are like the rest of the women I know: if they're intellectual they become aggressive; if not, they become boring.
7. *"Attend a leadership school."* How do all the leaders ever get their work done? Do you find them all combining last year's crop at the end of May?
8. *"Trip to Ontario to sample DDT-free wine."* A federal researcher at Vineland research station says this phenomenon occurs since, during fermentation, DDT decomposes into other non-toxic compounds. Most pre-1969 wines in Ontario were made of DDT-treated grapes. DDT-free wine may be a substitute for fluoride-treated water.

9. *"A new kind of hybrid near Suffield."* It's my understanding from the Ribstone Rustler that a rancher near there has developed a big brown bovine breed called Beaujolais. A unique feature is the 4% butterfat content of its milk is replaced by 4% alcohol. When crossed with Welsh Black you get a Mexican-style fighting bull.
10. *"Field trip to Peace River country."* This is in response to suggestion from the owner of a pizza joint who figures those Chianini cattle from Italy will solve the problem of uncertain water supplies. With their nice long legs they can walk 25 miles to the river every day for a drink.
11. *"Junket to Whitefish, Montana."* This trip is necessary to study what causes cows to run around with corners bit out of those plastic ear tags. Is it that they are suffering a feeding deficiency or do people use them to patch holes in ski boots?

I think I shall have a lot of fun tracking down these stories. I hope the reader has had as much fun looking over the list as I have had in compiling it!

Opinions regarding just what qualifies as being funny vary between people and change over time. Nevertheless, the human desire or, to put it more strongly, the human need for humor endures. Humor is a powerful emotion that often tips the balance of our internal decision making mechanism and it provides invaluable assistance to our memory. Without emotions one option is much like any other. It is our emotions that account for differences; they are responsible for making things matter to us.

HOW IT USED TO BE
– *The Calgary Herald* – October 30, 1971

If people over 44 have any reservations about Halloween today it is that kids don't use their imagination but are encouraged to go begging for sweets. Of course, the passing of the horse and buggy and the outdoor privy has eliminated most of the Halloween legends of the rural areas. It is the "outhouse humour" that is remembered when farmers get together to recount pranks of yesteryear.

One of the most imaginative pranks I recall was when a group of farm youths spirited away a privy to an overhead railway bridge. They waited till a freight train came along and dumped it onto a gondola car. The building travelled 50 miles that night and caused the freight department no end of grief figuring out a tariff on it. It caused the bereft family no end of grief, too, because they were poor and couldn't afford a new one. This stroke of genius was followed next year with six sturdy outhouses lined up in front of the Bank of Commerce the morning after with a sign imperially claiming: "Make your deposits here."

Ask a staid bank president, or a tough company president or a stern-visaged magistrate and you will find the imp of the Halloween prankster of 1936. There's a magistrate in Red Deer who is said to withdraw from all cases involving Halloween "vandalism" these days. Said magistrate was the ringleader of a gang who put a pig in the school in his home town in far-off Ontario. The students all got a holiday next day while the janitor cleaned up the school. This magistrate is said to be unable to keep a straight face or conceive a proper sentence.

There is also the charge by the middle-aged that kids are lazy today and won't work till dawn taking a buggy apart and reassembling it on a barn roof or putting a horse in the haymow. Their point is hardly valid today: What can a gang of kids do to indoor plumbing or where do you find a buggy? Sad to say, the familiar shout of: "Apples, cake or candy—Or over goes your shanty" is not heard in the rural areas these days.

Schmidt owes a huge intellectual debt to his many venerable associates and various other personages from whom he has acquired much wisdom for the benefit his readership.

PRESS IS PLUCKED – *The Calgary Herald* – August 21, 1973

Several months ago an uproar broke out in Italy when it was discovered two journalists had published a book based on a "bugged" church confession box. Keeping in mind that confession is good for the soul and that not even the agricultural congregation should be excluded, I loaned my black hat to my good and trusted friend, Thaddeus Xavier McMurphyvisk, and sent him forth with a 10-speed bike and cutty stool to survey the spiritual health of agriculture. Here is a running account of his work (McMurphyvisk is still running for the hills) as the result of bugging milking sheds, haystacks, farm meetings, box socials and field days.

A.M. Boswell and J.H. Berry: Forgive us, dear Thaddeus Xavier, for the terrible transgression we made in the summer outlook for the Canada department of agriculture economics division. We said top steer prices at Toronto may weaken slightly this summer from late May levels of $47 per cwt. Woe. Woe. The price went up to $57.

The Canadian Broiler Council: Intercede on our behalf, Thaddeus, for we have lost our humility. We chickenpluckers had a humble beginning in Saskatchewan—and the press was most welcome at our meetings because farm editors could help us. But now under the Canadian Broiler Council we have gone big time and make jet trips around the country. We treat our former friends shoddily and leave them outside the door. Remorse. Remorse.

A potato grower from Vauxhall: I missed church every Sunday last year because I was in the storage room bagging cull potatoes to sell in the stores, even though the potatoes had been contracted to a processing plant at half the price. Have I sinned, St. Thaddeus, because I owed the bank $75,000 and the company had thousands of pounds of unsaleable processed potatoes on hand. Or am I merely confused?

A grain grower near Grande Prairie: I humbly beg forgiveness, father McMurphyvisk, for including that 65-acre lake in my wheat board permit to collect a LIFT payment in 1970. But I will pay it back to the receiver-general of Canada. I will atone for my mistake by applying for a LIP grant to shoot ducks on the lake to give to the poor. Where do you want them delivered?

A flaxseed grower at Blackie: Dear, kind Thaddeus Xavier McMurphyvisk, I confess to breaking my contract to sell my flax crop at $3.40 a bushel. I had my eyes blinded with fury when I went into a supermarket and saw clerks marking up food which was already priced on the shelves. Is the sin of wanting $11 a bushel for my flax worse than that of the owner of the supermarket? I need your help on this one, Lord!

Health Minister John Munro: When I banned the use of stilbestrol (DES) for cattle feeding last January the price of cattle was 42 cents a pound on the hoof.

I confess there was no real need for banning this growth hormone, which allowed feeders to finish cattle faster. They said the beef price would go up to consumers as the result. I contritely admit I could not foresee the price of cattle going up to 57 cents on the hoof.

A plant industry division man: I heartily recant my act of meanness to my brothers in the marketing division of the Alberta department of agriculture. I brought out that report on projected heavy infestations of Bertha army worms much against their wishes. Next year we won't fund entomological surveys; we'll let the farmers guess where the armyworms are likely to be serious. Please, Thaddeus, teach me more about brotherly love.

Prime Minister Trudeau: Mea culpa. Mea culpa. My government tried to force bargaining power onto the Canadian cattlemen via Bill C-176. I thought they were wrong to resist this beneficial government measure. But, mea culpa, mea culpa, how wrong I was. I would never have been able to force export controls on cattle and beef if the cattlemen had already been given "bargaining power." Teach me, kind Thaddeus Xavier, the wonderful ways of the Western cattleman.

An Alberta egg producer: I have sinned against the Alberta Egg Marketing Board as I have 7,000 chickens and have not been paying the selling fee to the marketing board. Intercede on my behalf, McMurphyvisk, for I didn't truly think it would take the egg board four years to find me and my flock of 7,000. I am one of the meeker sheep in the flock and do not like to crow aloud about my business. I waited for their tax collector, who never came.

J.E. Duncan, president of B.C. Tree Fruits Ltd.: Hear the pleadings of a penitent who made an unjustified attack on the agricultural editor of The Herald, charging him to his editor-in-chief with misrepresentations and bias in April 17-18-19, 1973 columns on B.C. fruit marketing. Yea, his words were nothing compared to the actions of the dissidents about whom he wrote and, incidentally, gauged their temperament accurately. I'll atone for the attack by interviewing the growers personally myself the next time to hear their ideas with my own ears.

Stories mirror life insofar as they have a beginning and an ending; with a plot line in between the two definitive ends. The trick for all authors is deciding whether their mirror ought to be concave or convex. In other words, should the pivotal event be expanded or condensed to get the point across to the reader. Of course, the key to ending a story successfully is deciding on just what constitutes the main point to story. Retelling real life events is considerably more difficult because people continue doing one fascinating thing after another.

Schmidt kept a special file of hilarious, off-beat stories that could only originate in the rural areas. On slow days he'd pour a bushel or two into his column—then head back to the country for a refill. And speaking of stories, apparently speakers were never invited back to a farm meeting unless they could tell a good tale at the start of their speech. The end of story telling, according to Schmidt, came in 1987 with the introduction of P.C. (political correctness).

WHERE SHOULD A STORY END?
– *The Calgary Herald* – January 4, 1974

The December 17, 1974 column on the missing Bassano tape brings to mind another Bassano story, which shows this is an Alberta town where the unexpected is commonplace.

A young farmer and his young wife drove into town one day to buy some house furnishings. They loaded into the back of the pick-up, among other things, a piano, box spring and mattress. Having finished their shopping, they dropped into the local pub for some refreshments before proceeding home. There they met some convivial friends who persuaded them it would be in their best interests to stay in town for the dance that night. They left their truck parked outside the dance hall and proceeded to enjoy themselves until a point in the evening at which they had a fight on the floor and the young wife ran out in tearful fury, jumped into the truck and drove home over 15 miles of rough road at high speed. When she reached home she discovered a young man and his girlfriend in the back of the truck—and they hadn't gone there to play the piano.

Whether the story should end there is a difficult matter to decide. No doubt the reader will raise some questions—as he sits back in the comfort of his living room to contemplate the drama. How did the couple in the back of the truck get back to town, for instance? What finally happened to the young husband who was deserted on the dance floor? What was the reaction of the young wife when she discovered the stowaways after the wild ride? In fact, what approach did the scared stowaways use to attract the attention of the furious young wife?

Leaving it to the imagination of the readers to conjure up more speculation about this interesting bit of lore from Bassano, I now pass on to the town of Didsbury for another bit of lore, which has been passed from mouth to mouth at farm meetings for the last year. This story is about a farmer who was riding a horse serenely across a field one afternoon when all of a sudden the horse was shot out from underneath him. It seems a doctor had contracted an undiagnosed case of buck fever and had mistaken the horse for a moose. Once again the reader is left to his imagination as to how a man could be riding a moose. Would this be the only man to have his horse shot out from under him in Canada in that particular year? Would the hunter lose his licence—to practise medicine, that is? This story has been kept fairly quiet for many months to protect the next of kin of the deceased, I presume.

I've heard another horse story several times, which is supposed to be based on fact. A farmer near Stettler had an old horse which he loved dearly and which was too old to work. He didn't have the heart to destroy it so he asked a friend from Calgary to do the job for him next time he came out to hunt. The friend duly showed up accompanied by a companion. Coming to the back pasture where the horse was located, the Calgary hunter, a bit of a practical joker, said: "There's a horse I've been wanting to shoot for a. long time." Where upon he shot the animal dead. His companion became so incensed by the seemingly cruel act, he felled hunter with a blow from the butt end of his gun.

Almost everyone in the rural areas knows a story about rustling these days. The high price of meat invites more and more livestock thefts. However, grain prices are also high and this makes crops a target, although somewhat more difficult to steal. A Saskatchewan farmer, to his sorrow, found when he went to combine a 100-acre field of flax in a field some distance from home that somebody else had beat him to the job. Flax was then worth $10 a bushel and his crop ran 10 bushels to the acre. A most imaginative theft, to say the least.

It takes a great deal of imagination to accept the fact a fireman recently had to pay to go and put out a fire for another party. Around my hometown in Ontario they are still chortling at the avariciousness of a thick-headed gatekeeper at the Grand River Conservation Authority's Pinehurst Park. A fire broke out there and a call was put in to the fire brigade. As happens frequently in the case of volunteer fire brigades which make rural runs, the first few firemen on scene drove the pumper to the fire. One fireman arrived late and followed the fire truck in his private car. Arriving at the park entrance, he was told by the gatekeeper he would have to pay $1.75 admission charge. Nothing could persuade the gatekeeper he was a volunteer fireman. There has been some discussion about issuing the firemen with business cards.

Fascinatingly, words at times appear to take on traits similar to the personalities of people. And, not unlike the personalities of some people, words can, at times, rub people the wrong way.

RAPESEED CRISIS CENTRE
– *The Calgary Herald* – February 27, 1974

News headline—Rape Crisis Centre Opens in Toronto
News item—Rapeseed Association of Canada Meets in Calgary Tomorrow

Dear Rapeseed Crisis Centre: In my opinion Prairie farmers made an error in judgment in voting to continue the open marketing pricing of rapeseed. However, I am confident the time will come when farmers will vote for orderly marketing of rape. What can we do to avert a crisis until then? (Sgd.) Roy Atkinson, President, National Farmers Union.

Dear Rapeseed Association of Canada: Yesterday I asked my teacher what rape is. She told me you call it the Cinderella Crop. Therefore will I find the story of Cinderella under "fairy stories" in Calgary Public Library? (Sgd.) Inez Schmidt, age 13.

Dear Rape Crisis Centre: Last September I took my rapeseed to the (bankrupt) Diversified Crops Ltd. warehouse in Calgary. The door was locked and nobody was around. Should I have left it there anyway? (Sgd.) Thaddeus Xavier McMurphyvisk. Lucky Strike, Alberta.

Dear Rapeseed Association of Canada: I learned recently you had changed the name "rapeseed oil" to "Canbra oil." Your reason was to make it a more appealing product to the consumer. Using a reverse twist to reduce the increasing incidence of rape (23% in Toronto), do you think changing the name to a more explicit description of the crime, "sexual grand larceny," would be the right thing? (Sgd.) Ms. J. Clark, home economist.

Dear Rape Crisis Centre: We bought some rapeseed from the Saskatchewan Wheat Pool and found the erucic acid content was higher than the 5% allowed by Health Minister Marc Lalonde. We have had to substitute American soybeans. We are now fearful Mr. Lalonde may try to keep out American soybeans as well as American football. What should we do? (Sgd.) J.J. Banfield, Canbra Foods Ltd., Lethbridge.

Dear Rapeseed Association of Canada: Do you think the painting, the Rape of the Sabine Maidens, is pornographic? (Sgd.) Sgt-Maj. Louie Blooie, Balzac police morality squad.

Dear Rape Crisis Centre: We were told by the federal department of agriculture if we didn't grow Span low-erucic acid rapeseed we would cause heart attacks in rats. Since Alberta is a rat-free province, could you tell us where to acquire a supply of rats? (Sgd) Dr. C.O.N. Buller and Dr. B.U.L. Conner, horse physicians and surgeons.

Dear Rapeseed Association of Canada: The people are confused (and somewhat amused) by the similarity of our names and aims. Could you please arrange to send a representative to Buck Valli's sheep fold for an exchange of pleasantries and mail? (Sgd.) Carolyn Bode, Rape Crisis Centre, Toronto.

Change permeates, in fact, defines post-modern life, whether it is in terms of defining what is humorous or redefining the definition of traditional and commonly used words. Nevertheless, there are those who do what they can to resist or at least attempt to retard the pace of change. Schmidt provides a glimpse of two such people who have chosen a much less "changeful" lifestyle.

LEMONADE FARM PROPRIETOR
- *The Calgary Herald* - September 6, 1974

It's only two leagues east of the grain elevators at Pulteney to the Lemonade Farm where Eddie Bakos and Alkali Bronson preside over 27 steers, a Hereford cow named Nellie, three cats, a dog, two horses and 75,378 hungry ducks which would clean the wheat out of all the fields if permitted. There is no chance either of them will become corrupted by some of the useless and expensive luxuries of modern urban living. "Things are primitive—but I like them that way," said Bakos, who is proprietor of Lemonade Farm. It is not Grandma Moses primitive but Sodbuster primitive—a primitiveness dedicated to growing flax, barley, wheat and a vegetable garden. In this garden, I helped can a few peas over the weekend by the quick-eat method. Um-m-m, good! The Swiss chard and the corn were top rate, too.

There are a couple of concessions to modern-day soft living. One is a tractor with an air-conditioned cab. However, should the energy crisis or built-in obsolescence put it out of commission, there are three sets of harness in the bunkhouse. One may have to step gingerly over a dozen empties to reach them but this back-up source of energy is there in reserve. Another is a bottle of lemonade made from pre-mix crystals from the supermarket and mixed with alkali well water. It's kept in the refrigerator. It can be used to cure many dolorous maladies but if it fails it can be mixed with other liquid potions. Pull out the bread drawer north of the kitchen sink and one may be lucky enough to find a 26-ouncer there—not necessarily a 26-ounce loaf of rye.

Life is geared to the slogan on the refrigerator door: "When life gives you a lemon—make lemonade." Alkali Bronson was not hired. He adopted Lemonade Farm last Easter. He came cruising in from somewhere in the Cariboo country of B.C. to park his trailer in a safe place in the yard. He liked the look of the place and there weren't likely to be too many women coming around (Mrs. Bakos is a nurse at the hospital in town and lives in the Bakos town house). He looked over the place and found many attributes. In a display of agrological know-how, Bakos pointed out one of the strong points of the farmstead: "Look at the barn. It's 53 years old and hasn't got a sway in it anywhere," he said.

Alkali took this into consideration, along with a few other things. He adopted the farm and decided to go to work. He reserves the right as a free and independent man to take the day off now and then and he has a trap line established through the nearby towns of Nanton, Granum, Champion, Nobleford and Claresholm. He likes primitive living, too. I mean you won't find a bathroom at Lemonade Farm. On the first visit I made, it was necessary to wend my way past the repair shop to a stately two-hole edifice surrounded by bales of straw. The edifice had a lean on it. The bales served to prevent it from going completely askew and rendered it secure from the winter winds. The ancient structure has since been replaced by a more stately structure. There is the possibility the user may soon have more protection from the fall frosts if the seat is covered with a calf-skin. There was no promise imputed in the deliberations about it, however.

Where there is a refrigerator there must be electricity. The kitchen is lit with only a single bulb. However, as a back-up source of light there are five lanterns, one of which is a switchman's lantern carried by his late father, who was latterly a C.P. Rail conductor out of Medicine Hat on the Spokane Flyer between the two cities. Bakos regrets the Spokane Flyer wasn't revived for the World's Fair in the Washington State city this year. He felt it would have brought back old memories like hog-killing time on the farm, which used to occupy several days; and trips into the hills to pick berries for preserving.

The Lemonade Farm could well become a national institution if the fiscal campaign by Eddie Bakos and Alkali Bronson takes hold. They are firmly convinced this country should go off the gold standard and onto the manure standard. Food is of prime importance. It takes fertilizer to grow food. Food can't be grown with gold.

Personally, I am happy that the Lemonade Farm is primitive enough that along with Nellie, the Hereford cow, there is a hand-turned cream separator still in actual service. I was able to revive my hand-milking technique of teenage days and, pushing aside her hungry calf, was able to extract a quantity of milk and separate some cream to mix with a beaker of Lamb's Navy rum. The cream separator is preserved in pristine cleanliness. They don't make cream separators any more. That's an indication of the sad state which the dairy industry finds itself and why we're short of butter in this country. Yes, a visit to the Lemonade Farm can provide a pleasant interlude to a former barefoot boy who finds life in the city leaves him soft, flabby and disaffected.

Superficial changes notwithstanding, there may be something reassuring in the realization that at the most fundamental level human activities, despite the enormous possibilities for different twists and turns, remain essentially unchanged in their underlying propose.

MUCH ABOVE BULLHORN LEVEL
- *The Calgary Herald* - November 14, 1974

Last week readers like myself, who learned the basics of sex around the barnyard, were treated to a refresher course in the pages of *The Herald.* A panel of speakers including: a prostitute, sociologists, anti-marriage advocates, professors, gays and religious spokesmen were presented on the platform of a human sexuality seminar at the University of Calgary. They brought the whole issue—whatever it was the university was trying to prove—down to the barnyard level. There was nothing left to the imagination and nothing left to discover tomorrow. Not even with your favourite heifer or the old sow.

This brings me around to a man who knows his way around the barnyard, H. Gordon Green. His latest book, *Diary of a Dirty Old Man,* is about sex. It is published by McClelland and Stewart and sells at the university bookstore for $7.95. This book is a switch from the prurient, bumbling U of C types, who use the dung fork to pitch sex at the public. Green's book takes sex from the barnyard to the normal level of discussion we expect from the university—taste, mystery and sensitivity—where it really belongs. Green is a farm writing friend of mine who is best known to CBC listeners as the Old Cynic. He was on the staff of The Family Herald of Montreal. When it ceased publication he reassumed his career as an English teacher at Dawson College, Montreal. He lived with his family on a 332-acre beef-breeding farm at nearby Ormstown.

I have known a number of middle-aged men who shucked off their middle-aged wives to cohabit with women as young as their daughters. In the interests of sexual equality I should restate this. I have known a number of young women who have taken middle-aged men away from their middle-aged wives. Few, if any, have had the nerve to write down the story of what society generally regards as sexual lunacy. While he passes off his diary as primarily a work of fiction, there is more truth evident.

Green describes on a day-by-day basis every move he has made from the day he received an anonymous love poem from a Grade 13 teenager until the time she moved onto the farm with him. He was 55 when he received the billet doux. The move was accomplished with the full knowledge of his wife and family as there was an old stone house on the property that he used as an office and refuge from what may be described as a withered marriage. It's hard to determine what has gone wrong with it: Whether she is too sanctimonious to accept his foibles (which are considerable) or whether she feels that after age 50 she is entitled to practise frigidity. Green never mentions sex with her as he does with a flock of other chicks. During the big "scene" when wife Edith finally realized she had lost her husband to teen-age Sherry, she called him a "fornicating old fool." But my sympathies weren't with her—mostly because Green had done a good public relations job for all dirty, but lovable, old men, but partly because she seemed responsible for the marriage deterioration. Her attitude toward this dirty old man now seemed to be "if I can't have his body and soul, which I don't really want, I am going to get all his money so he won't be able to lavish it on that Other Woman."

As a background to the exposition of his love life, Gordon Green skilfully works into the mix barnyard jokes, graffiti, some of the scripts he has used on the CBC, and his own philosophy of life. The diary may shock the properly pious. However, it is done in simple, understandable prose which raises it much above leering, pimply-faced, dope-crazed adolescents trumpeting sexual attitudes into a bullhorn at the University of Calgary. That old fogey, Green, has shown those blasé urban super-sophisticates (who figure they have learned all there is to know about sex by Grade 6) a trick or two. The best of his bag of tricks is to be able to sit back on his farm and put his experience on a hauntingly beautiful level with his pen. It is much more effective than assaulting the public's sensibilities on a sideshow basis. Green's last word is that Sherry is still with him.

Some are gifted with the facility to say a great deal with words, others communicate a good deal with no words. Schmidt provides examples of what distinguishes both types.

TOO SHY TO TALK - *The Calgary Herald* - March 12, 1975

Even before he assumes office on April Fool's Day, L. Denis Hudon, new deputy minister of agriculture, has been put through the wringer by critics. The main thrust of the critics is that while Mr. Hudon has impressive credentials, he may not know anything about agriculture. Some say he doesn't want to know anything about Canada's agricultural policies and programs. To me, this isn't a valid concern. I mean, who in hell does know anything about agriculture today? The answer: Only a few rich farmers. They got that way by doing exactly opposite what the federal agriculture department told them. So there is hope—with Mr. Hudon in the department's most sensitive job. If he doesn't know anything about farming and farmers listen to what he says they'll all have a chance of growing rich.

On the other hand the new deputy is a shy man. He may have enough sense not to tell farmers anything, thus reinstating the role of government to normal. Then nobody will be able to pin anything on him. He is so shy he has been reluctant to talk to reporters until he assumes office. I feel the organized farm writers could have bridged the gap by inviting the new deputy minister to Alberta for the week of the Calgary bull sale, pulling hay out of a rancher's stack near Coronation, inspecting an artificial insemination unit or trying to haul grain to plugged elevators. Of course, there could be a drawback. Hudon could be like that schoolboy who complained after a lecture that he now knew more about penguins than he wished to know.

Some of the criticism over the Hudon appointment was that S.B.Williams, retiring deputy, was not informed of the appointment. He read it in the paper first. Eugene Whelan was a bit snappish over questions about the appointment. He explained that while he was consulted about it, Prime Minister Trudeau had the final say. When asked by Gordon Towers (PC—Red Deer) for the basis of selection (there were four assistant deputy ministers all eligible), Mr. Whelan told him it was none of his business. The prime minister is using the collegiums approach to shake seven government departments loose from the grip of deputy ministers. And why not? What's wrong with the prime minister using the *Time Magazine* approach of assigning jobs? Even though Trudeau himself doesn't know anything about agriculture, he marched over to Great Britain and got back that cheese contract we lost a couple of years ago. That's despite the fact he doesn't even know how to make cheese.

Word is that Mr. Hudon keeps a clean desk. If fastidiousness is a merit, he may make a good deputy. On his way through the finance department, Canadian International Development Agency and as deputy secretary to cabinet for operations, he was known as a good administrator. Whether he can come to grips with an agricultural civil service, which has proliferated to more than 10,000, and still keep a clean desk may be a hat trick difficult to perform. However, there is talk of him shuffling off some of this hired help—some to health and welfare department, which is charged with border post inspections. Those antiseptic footbaths against entry of livestock disease may finally be installed in international airports if this happens. It would be no loss to ship to consumer affairs those women in the agriculture department who are always advising consumers what food is cheap. The business of the agriculture department is to tell farmers how to take more out of consumers' pockets.

It occurs to me farm writers could offer public figures a greater appreciation of agriculture than they do. I referred earlier to a public service we could have done by inviting Mr. Hudon to Alberta for an agrarian primer course. I think, for instance, we could render similar service to our contemporaries in the sports writing field by providing them with more fruitful visits to farms. Their activities in regard to farms are largely confined to writing about horses and baseball farm clubs. Few ever visit a farm to see the boars and bulls, which are equally important.

There are two sports reporters in particular I believe the farm writers could help, this being International Women's Year. They are Robin Herman of The New York Times and Marcelle St. Cyr of a Montreal radio station. These are two female sports writers who have been used by several National Hockey League coaches to garner some free publicity by barring them from locker rooms after the games. The women say they want the same right to interview players as male sports reporters from competing papers.

As an aside, several Calgary sports writers are ready to kick a well-known weekly editor high in the jockstrap for suggesting the reason male sports reporters visit locker rooms may be the result of latent homosexual tendencies. Rather than offer these two female sports reporters a roomful of half-naked men, the farm writers can offer them 300 head of naked exotic bulls at the Canadian Western stock show and sale at Edmonton Exhibition grounds March 22-29, 1975. Girls, there'd be nothing more exciting than observing the libido of a 1 - ton Chianina bull in the showers after a hard battle in the show ring at Edmonton. The Canadian Chianina Association has a slogan: "Chianina bulls are better lovers."

After the word meisters, Schmidt and Whelan, have one of their many sizzling exchanges, Schmidt fires off his most potent missile – humour.

FAMILY FARMS WILL DISAPPEAR
– *The Calgary Herald* – April 11, 1975

I've just had a of a bit of a go-around with Agriculture Minister Eugene Whelan, who feels I am a nasty old man for that March 12, 1975 column on Denis Hudon, the new deputy minister of agriculture for Canada. I have replied to Mr. Whelan to suggest it isn't all my fault that I have developed into a schizophrenic, unable to determine whether I should be certified insane as a farm writer or jailed for treason as a political writer. I humbly suggested that in his eyes I may not be suited to writing the type of politics that has become increasingly common in agricultural columns. However, I felt compelled to add that were it not for the fact that party politicians had increasingly infiltrated the agricultural field, he would have no grounds for complaint.

So I can shuck my split personality and ensure my return to sanity and my desired station in life as a self-taught political writer without benefit of exposure to all those American university professors, I have developed the:

Ten New Commandments for Farmers:

1. Farmers can do nothing politically for themselves. A strong, charismatic leader is the answer to all problems.
2. Farmers should indulge in economic class warfare and keep pesky city consumers off marketing boards.
3. Each farm organization needs a constitution enabling it to pass resolutions sure to be pigeon-holed by politicians.
4. A large staff of farmers who are too lazy to farm should be hired to see if they will do any work for a farm organization.
5. Farmers should convince themselves that if the government is made up of "baddies," the opposition must be made up of "goodies." (Two elections ago the Alberta Socreds were the "baddies." Now there isn't enough of them to be considered "goodies.")
6. Even though they pass NDP legislation, politicians who speak of the "evils of socialism" and "income stabilization assurance" must be "goodies."
7. When things get bad, organize big rallies and invite speakers to tell those farmers assembled things are getting worse.
8. Build an image—not "left" or "right"—but stay in the "middle of the road" (where it's easier to get knocked over).
9. If by chance politicians make election promises and break them after the election, they should be forgiven. Politicians are entitled to their shortcomings the same as everyone else.
10. By observing the first nine commandments farmers can ensure the family farm will soon be a thing of the past. They should plan to sell their farms to the government landbank or get on the nearest government payroll so I, as a political commentator, can write fulsome paeans of praise about their good works.

One of Schmidt's most knowledgeable alter egos—who's breadth of expertise embraces all that is to known to humankind—is ever-ready to provide in-depth analysis of any situation.

THE RIGHT TO GO BROKE
– *The Calgary Herald* – May 15, 1975

I spotted Thaddeus Z. McMurphyvisk leading a quarter horse filly away from the Calgary spring horse show one night last week. "Nice filly you've bought," I observed.

"Yep. She's going to be my wheel-horse for the Canadian Wagon Train, Eastbound Division," he replied. "We plan to leave the Calgary Stewpede on July 4, 1975 to make a 2,000-mile covered-wagon trip to Kleinburg, Ontario."

"Kleinburg?"

"Yes, those people from Kleinburg on the Canadian Wagon Train coming West have never been to the Peace River country nor been bitten by the mosquitoes there. I have never been to Kleinburg and been bitten by a bluebottle fly, either."

"I won't ask you why you want to go back," I said.

"There are basically two reasons," he replied, cutting off a chunk of Old Goathoof chewing tobacco. One was that speech of Cathy Philip, a University of Calgary resources and environment professor, to the centennial Frontier Calgary conference. They quoted her as saying 'some pioneer women didn't stop crying for three months after they hit Calgary.' So I'm going to round up the men who had to put up with them and take 'em back to Ontario. Admittedly many are pretty old but they deserve something better in life than this."

"Most women are crying or whining about something most of the time."

"You bet," McMurphyvisk spat. "And the other reason I'm going is to start this wagon train East to escape that lunatic fringe we're inheriting from the East. There was this dude named Thomas Berwager who, last January, wrote to a friend of mine in the Alberta department of agriculture in Edmonton. Said he'd just spent four years at Penn State University in rural sociology courses to prepare himself for 'the opportunity to learn and grow in the way of life found in the northern areas of Alberta.' He says he and his wife have limited farm experience but hope to develop further. In this respect he considers himself adequate at outdoor survival and hiking."

"A lot of farmers are hiking away from bankruptcies these days," I pointed out.

"With all these weeping women and experts in survival from the East, we just want to get away from it all in Kleinburg. We'll only have Pierre Berton to contend with there. And anyway I'm sick of trying to live with the provisions of the Alberta Coarse Grain Marketing Control Act and the Billiard Rooms Act," he pouted.

"The regulations are almost as unreasonable as the Women's Institute Act," I suggested. "By the way, what's your schedule?"

"We aim to be in Ontario to save Bill Davis' government from being defeated and having the province revert back to a Mitch Hepburn-type of government. You remember that fascist onion grower? They say some guy named Nixon is out to get Davis."

"As long as he isn't the Nixon from San Clemente..." I volunteered.

"No, but guys named Nixon are suspect in politics today."

"This trek will be just like jumping off the edge of the earth."

"Yeah, but there will be some compensations going back East. We're going to take a few Western steers and pick up some feed grain and feed them out in Kleinburg and make more money than we can feeding them here. Our wagon train is going to apply for the feed grain transportation subvention."

"Quick thinking. But you might need them if you run out of food on the way if the price of beef goes up again?"

"No. We plan to have more exotic fare than beef. You will recall the Donner expedition ended up in disaster by eating each other. Well, we plan to cut off one half the group in the Canadian Wagon Train coming West on the north shore of Lake Superior and eat them," chortled McMurphyvisk.

"I never figured you for a bloodthirsty character. What other qualifications do you have for leading the Canadian Wagon Train, Eastward Division?" I demanded.

"I have just been sitting through some hearings of the Maxwell Mackenzie commission of inquiry into beef prices in Edmonton and listening to those farmers from the Peace River country. They have been telling the commissioners they are losing up to $265 a head on cattle all winter. Now these people in the Canadian Wagon Train, Westward Division, are going to the Peace River country in Alberta to start a two-section beef ranch. They would be better dead than going into a situation like that," said Thaddeus.

"But," I protested, "You are beginning to talk like Stalin. You want to keep people away from Alberta before they have the opportunity to go broke. That ought to be their privilege."

"Not with Ontario caught in a worse bout of inflation than Alberta is suffering at the present time," he said as he strode resolutely into the sunrise.

While Mr. McMurphyvisk's observations may raise an eye-brow or two, the technology that goes into the new farm equipment is truly eye-popping.

TIME MACHINE INTRODUCED
– *The Calgary Herald* – June 26, 1975

RED DEER—Martha (answering the telephone): "No, George, dear John isn't here. He's out drinking again, driving that new John Deere four-wheel-drive tractor of his on the summer-fallow."

George: "You mean . . ?"

Martha: "Yes, George, dear John never comes home at night any more since he spent $48,000 on that tractor and put in a beer cooler as optional equipment."

George: "You must be terribly lonely."

Martha: "Yes. That John has also put air conditioning in the noise-proof cab and he has a radio. He sits there on that new posture seat with Uncle Ben (beer), listening to CKUA playing those Beethoven quartets." (Weeps.)

George: "And there you are slaving over a hot stove with nothing to drink but coffee becoming a regular tractor widow."

Martha: "Dear John would never notice. He's so in love with that big green machine and its turbo-charged 275-horsepower engine, the tilt-telescope steering wheel, the seven lights and a spotlight on the cab and 590-litre gas tank that he's got no time for me any more. Why, next Sunday he's likely to drive it to church."

George: "Martha, if I promise never to buy a John Deere four-wheel drive, will you run away with me?"

Martha: "Yes, George, as long as you can go faster than 20 miles an hour. John won't be able to catch us with that tractor of his. I'll move out tonight."

These thoughts were running through my mind Wednesday as John Deere Limited representatives turned me loose with one of their new behemoths and a 28-foot disk on the farm of Jim Morrisroe at Red Deer. The last tractor of that make I had driven was a 19-horsepower model D John Deere in the 1940s. The new tractors are the difference between riding a jet and a jitney. The company didn't invite just anybody in for the week-long demonstration at Morrisroe's. They picked 600 Alberta farmers with four sections or more who could make use of the new "time machines." Gordon Milne of Calgary, company area supervisor, says they're dubbed "time machines" because with one such time-saver one man can run at least 2,400 acres by himself.

Thoroughly modern farming extends its cultivation techniques to plow the vertical rather than horizontal, but nevertheless, fertile pay dirt found on the silver screen.

FARMING IN THE ARTS
– *The Calgary Herald* – January 15, 1976

Recently I have been entertained by some movies in which I know most of the "actors" on a first-name basis. There were a great many familiar scenes, including Heritage Park footage, of early methods of harvesting and delivering grain in *The Way It Was*. This is a documentary produced by the Alberta Wheat Pool, directed by Bob Willis of Canawest-Master Films of Calgary. Boyd Cuthbertson of Okotoks is convincing in his observation that it takes more than $2,000 and a small piece of land to get started in farming these days. This is the message the agriculture department wanted to get across to urban viewers in the National Film Board of Canada effort called: *Following the Plough*.

The National Film Board has made another short for the Canada department of regional economic expansion (DREE). It is called *The Eight-Mile Haul*, and is emblematic of the measure, in terms of time and distance, it took a farmer with a team and grain tank to travel to town and back before supper in making a grain delivery. The team and wagon have gone and so have many other important things in the rural areas. DREE's story is that many people still want to live in the rural areas—and DREE can make this possible for the people who do. Some footage of the Slave Lake development is shown.

Edmonton's Century II Motion Pictures Limited has produced a film about irrigation possibilities, called *The Magic of Water*, for the Alberta department of agriculture. It was good enough that it was given an award at the annual educational aid awards of the American Society of Agricultural Engineers last summer at Davis, California. The film shows why Jay Purnell, director of the department's irrigation division, feels another four million acres should be brought under the ditch in Alberta.

The wheat pool's film will be used in a grain museum dedicated to the new agricultural complex of the Calgary Exhibition and Stampede by the pool delegates as a City of Calgary centennial project. If the Stampede board doesn't soon show some signs of building that agricultural complex, this may be the first time in Canadian history a museum will have to be built in advance of the fact. Incidentally, lack of an agricultural complex isn't going to worry the sponsors of the world Hereford show in July. They have been given the Stampede Corral to hold meetings and show cattle.

Most museums are rather seedy as far as I am concerned; and bar(e)ly able to finance themselves. However, I am assured by my wheat pool friends this will not be so at the grain museum; neither will it be corny. It was just a straw in the wind, but I asked if it would be run by a kernel. "No, by a sergeant," replied Gary Sargent, a pool researcher. He said he didn't mind me chaffing them about who would head it up, nor that I took a rye look at the whole deal. My concern was that it wouldn't be run by a bunch of flax (flacks) out to get a lot of free publicity.

There have been allegations that a great many things done by Wheat Board Minister Otto Lang have gone against the grain of many grain growers. A movement may be afoot to stuff him like a sack of flour and put him in the grain museum. There is no truth, however, the Rape Crisis Centre will set up the only working display. This has been threshed out well in advance. The directors have plowed through the possibilities of placing the metric bushel on display. Which may bring up the question: What is a metric bushel?
10 grains of wheat equals 1 metric shovelful
1 metric shovelful equals 1 metric bushel
100 metric bushels equal 1 metric truckload
1,000 metric truckloads equal 1 metric elevator
1 metric elevator equals 100 metric boxcars
1,000 metric boxcars equals 1 metric terminal elevator
1 metric elevator holds 100 metric longshoremen
1,000 metric longshoremen equal 1 metric farmer
1 metric farm produces 1 metric boatload
1 metric boatload equals 100,000 metric co-ordinators
1 metric co-ordinator equals 100 metric grain commissioners
100 metric grain commissioners have 1 metric grain of brains
1 metric grain of brains equals 10 grains of wheat.

Schmidt's brand of humour is entertaining as well as informative and more will be provided later. Humour, like a good meal, needs a little time to be digested.

CHAPTER FOUR

Peripatetic Traveler Through The Gardens Of The World

AGRICULTURE AROUND THE GLOBE

Schmidt had not traveled outside of Canada prior to 1968. Then, in mid-winter, he was assigned to accompany a group of farmers and ranchers to Hong Kong, Australia, New Zealand and Japan. Subsequently, he has been out of Canada at least once a year almost every year since and to date he has accumulated passport visas from 43 countries. His first excursion set the tone for his approach in reporting on his trips abroad. Essentially, his method is to collect new ideas and research different methods for doing things to take back home to share with his readers.

Reading about the enigmas that Schmidt observes in other parts of the world proved extremely popular with both rural and urban folks. Being extraordinarily perceptive, he is able to write far more realistic descriptions about what is happening in other countries than the travel magazines or glossy promotional brochures. Whether it is to provide his readers with some insight prior to embarking on their own journeys, or to ignite the imagination of those who can only dream about going to exotic places, these columns are among his readers' favourites.

In retrospect, these columns contain valuable background material for understanding the people and events that have shaped today's world affairs. Subdivided into regions, the following collection, selected from the large number of this farm writer's travel columns, demonstrates his often uncanny foresight regarding the future impact of some of the agricultural developments he witnessed in various countries over time. Some of his views show remarkable insight, others are highly controversial, but all are interesting and thought provoking.

Most importantly, he encourages his readers to develop a global point of view as he underscores the interrelationship of all the natural phenomena that is inextricably intertwined with the interdependence of people everywhere. He explains that conservation of natural resources is not only a major concern all over the world, but it also offers potentially lucrative new business opportunities. In his articles, written years ago, he showed that protests against farm prices, taxes, globalization and free trade are far from being novel issues. Even so, international trade has long been of great benefit for Canadians, who produce far more food than can be consumed within its domestic market.

Spanning more than thirty years, this agricultural tour begins in the Far East and circles around the world to end up back in the Pacific Rim. Countries of special interest, particularly the former Union of Soviet Socialist Republics—that has an agricultural climate similar to Canada's but has exercised a far different political approach—are revisited several times to give broader view of some of their developments. While the differences he observed inspired Schmidt to write many fascinating stories, it is interesting to note that some of the problems faced by various countries are not only quite similar, but continue to challenge food producers still. Taken in totality, his writings provide a panoramic overview of humanities' primary and most important industry—agriculture.

FAR EAST - ORIENT

HONG KONG - *The Calgary Herald* - February 23, 1968

There are a great many dairy farming establishments in Australia and New Zealand but the most unusual and unlikely one I saw on the recently concluded Pacific tour was on the island of Hong Kong. Nearly 4,000,000 people are jammed into the 400 square miles that comprise the colony of Hong Kong, which is largely precipitous mountain terrain. (The mountains are so steep some of the people have been shoved off into the water; 140,000 of them make their homes on sampans and shacks over the harbor.) It is surprising, therefore, to find a 250-acre dairy farm perched atop a group of hills in the centre of the island. It's surprising, first of all to find the farmland hasn't been overrun by people, and secondly that crops can be harvested off the steep slopes. No Canadian dairy farmer would even think of attempting to farm hillsides that rise to 800 feet from the sea in less than a mile. However, the 1,800 cattle—Holstein and Ayrshire predominantly, with a few Jerseys and Guernseys thrown in—seem very contented in their surroundings. They don't have to forage on the steep hillsides on which a mountain goat would have trouble negotiating. The grass is cut by agile Chinese (mostly women) who carry it on their backs to the cattle. The milking herd is 1,200 and there are 600 head of young stock. They are housed in strongly built stone barns, which have a capacity of from 18 to 180 head. There is only one which houses 180 and it is a two-storey barn, the first one I have ever seen.

David Shaw, the Scottish-born farm manager for 16 years, said that while it would be more economical to have the milking herd in one barn, this was impossible because of the disease problem and the typhoon problem. He explained that because winds reach a velocity of 160 miles an hour in typhoons, it was necessary to build a heavy low stone structure with a roof which wouldn't lift and the barn itself wouldn't blow away. That is the principal reason the herd is stabled in small isolated units. (In the typhoon two years ago something like 12 inches of rain fell in four hours and one of the hillside barns was partially washed away and was isolated for three days.)

The other reason is that there are two or three outbreaks of foot-and-mouth disease annually. There is a better chance of controlling the disease in an isolated unit than if the cattle were all in one central barn, Mr. Shaw said. He has learned how to live with the disease.

I know somebody is going to protest I have my figures wrong when I suggest that a herd of 1,800 cattle can be maintained on less than 150 acres of grass—or one acre for six head. To a Prairie farmer this couldn't possibly be so.

However, Hong Kong is in a hot humid climate and the tropical grasses grow so fast you can almost see them shoot up. Napier and Guinea grass will produce five to six cuts of grass 4 to 4 1/2 feet high from April to October. Mr. Shaw said each acre will produce up to 70 tons of green matter each year. Of course, the secret of this rank growth is the 86-inch average rainfall each year and all the manure from the herd goes back onto the land. "We figure we do a materials handling job amounting to 25,000 tons a year," he said. "That is remarkable as it is all done by hand. Some of the grass is handled twice. The cows can't eat it all up in the April-October growing season so we ensile some 2,000 to 3,000 tons every year." Ensilage is made in stone silos built into the hillsides so that the tops of the silos are level with the ground for ease in filling.

The farm was established in 1886 by Sir Patrick Manson, the Scottish scientist who isolated the anopheles mosquito as a carrier of the bug which causes malaria. He imported 80 Holsteins from the Seattle-Vancouver area to supply milk for the British residents of the colony. Further importations were made from Australia, the United Kingdom, Holland and Canada.

When the Japanese captured Hong Kong in the Second World War, there were 2,000 head of cattle. When they left there were 230. An importation from Harry Hays of Calgary helped build up the herd again. There have been no Canadian imports since the Hays importation but Mr. Hays made a visit there 18 months ago, Mr. Shaw said.

TOKYO, JAPAN – *The Calgary Herald* - November 23, 1974

Last week I booked off on a holiday and in a matter of hours found myself pushing through the dense crowds at Haneda Airport in Tokyo. There were a number of friendly Alberta faces on the JAL (Japan Airlines) flight, as well as 50-cent drinks and a sampling of traditional Japanese food. The friendly faces belonged to a group of Alberta government officials and legislators, agricultural marketing board people and businessmen who formed a mission of 35 going to Japan to promote increased meat sales.

The cornerstone of the promotion was an Alberta meat festival in the main ballroom of the Tokyo Prince Hotel. For those who have been to Tokyo this hotel is beside the 1,100-foot Tokyo Tower. It is also a sister hotel to a posh hostelry which opened in Toronto last June. The simple progress of setting up the meat festival was harried by the fact President Gerald Ford was staying in the vicinity while paying the first state visit of any United States president to Japan. That situation was complicated by anti-Ford demonstrations of 100,000 people, a one-day railroad strike that turned Tokyo into a ghost town for the day and an earthquake which shook things up. Among the guests invited to the Alberta meat festival were officials of some of Japan's vast trading companies. At least six of these companies have a greater sales volume than General Motors. And speaking of sales volume, the Albertans had talks with Zennoh, the national federation of agricultural co-op associations, whose annual grand total of business is well over $9 billion annually. If comparisons are not invidious, the budget for the province of Alberta is only $2 billion.

Incidentally those railway workers were also protesting about the low Christmas bonus they had been offered. The custom of the country is to pay low and hand out Christmas bonuses of anywhere from $1,000 to $3,000 for most workers. Company executives operate the same way. Their pay may be nominal but their expense accounts —especially for entertaining clients—are fabulous. It seems to be a mark of honor to see who can lay on the best entertainment. All entertainment bills are paid by credit cards. The practice of the bistros to pad the bills and later split the "dividend" with the company executives doing the entertaining is winked at.

Those stories about high prices in Tokyo are all true. A trade coordinator with the Alberta Export Agency recalled a recent trip to Tokyo. While there he was entertained by a trading company executive at a geisha-type party where the drinks may run $50 apiece. He estimated the bill for six people that night was $3,400. This provides a bit of background to the big leagues in which the Alberta mission was playing on this occasion.

It's a big jump for Sten Berg from his hog farm at Sherwood Park to the offices of the giant Mitsui & Co to discuss forward contracting. Mr. Berg is a member of the Alberta Hog Producers Marketing Board marketing committee.

There was Howard Falkenberg, fresh off his broiler operation at Wetaskiwin, deep in conversation with a marketing man at a glittering reception sponsored by Nissho-Iwai on the 17th floor of its skyscraper. He is chairman of the Alberta Broiler Growers Marketing Board. And Tom Butterfield, who was riding a horse at his feedlot at Ponoka a couple of days ago, riding in a taxi to keep an appointment at the Nichimen and Co offices. He is vice-chairman of the Alberta Cattle Commission. They were making follow-up calls for a week after the meat festival. It's hard to know the effect of these calls. It's hard to know if the Japanese traders were discussing the impact of the Alberta promoters or the earthquake the previous Saturday.

For most of the delegation the November 16 earthquake in Tokyo was their first experience. At 8:32 a.m. I was on the point of getting up when the bed started rocking and the headboard hitting the wall. The whole room began shaking and creaking as if a giant hand was shuffling things around. It lasted about 10 seconds. For a minute afterward the curtains in the room on the 10th floor could be seen moving. Those on the ground floor didn't notice the tremor. Nobody panicked but one wondered what was the best course of action, especially with that tower next door which could tumble. There is no book of earthquake etiquette. The quake was serious enough, however, that the subway shut down for four minutes.

CHINA – *The Calgary Herald* – March 8, 1984

The January 12th column on the Western Canadian fertilizer industry passing up a chance to sell $7 million worth of anhydrous ammonia in the Pacific Rim market caught the eye of a Calgary China broker, Pino Leo, president of Invescana Consulting Group. His profession is a new one which has sprung up in Western Canada as we move to take advantage of new trading opportunities with China. The visits of Premier Peter Lougheed to China and of Premier Zhao Ziyang to Canada were merely window dressing. Trading with China is hard and difficult work.

Two extremes are apparent among Westerners and overseas Chinese alike (including those from Hong Kong) who trade or attempt to trade with China. Some are totally successful. Others are abject failures. There seems to be nothing between the two extremes. Those who can't make deals complain of too much time spent negotiating, too much red tape and orders from China not delivered on time.

The China broker said, "Most of these complaints and grievances arise, in part, from inadequate, or lack of understanding of trade barriers between China and the West— differences in customs, trade methods, politics, ideology and languages. Those who don't master these essentials are the persons who fail," he said. "Few realize what factors contribute to this failure. And they are: communication breakdown, lack of proper connections and inappropriate approach. These lead to frustrations, disappoint-ments, financial losses and eventual failure."

Those who are successful have realized that China was, is and will be, on the whole, run by one-man rule as opposed to democracy. From this one-man rule has emerged a common mentality, which is deeply entrenched in all Chinese government bodies. The mentality, which dominates the behavior of Chinese officialdom's way of doing business with outsiders, is totally different from Western styles and ideas we take for granted.

A great many Westerners and Chinese born and bred in Hong Kong have seen attempts at bribery cause deals to fall through. They have not become aware that today bribery is no longer workable and acceptable in China. If government officials discover bribery, the penalties are severe. The opening of the Chinese borders to foreign trading follows the Russian model. For years, the Communist government kept out foreign products until it was threatened with a revolt if more foreign "goodies" and food were not made available. Then the floodgates were opened.

One little-noticed phenomenon that Leo spotted on a four-month trip to Hong Kong is that there are numerous bona fide tenders for a variety of commodities and services that China is making known through non-official channels. Many transactions have been consummated, some worth millions of U.S. dollars. Another phenomenon, that is a popular misconception, is that those in the provincial and lower hierarchy of the political and administrative systems have no say in trade. This is not true. Deals can be and are made through those in lower hierarchies.

The Former UNION OF SOVIET SOCIALISTIC REPUBLICS

Schmidt's 1967 Canadian Centennial project was a trip to the North Pole. However, he didn't achieve his goal until 1971, when he flew from Fairbanks, Alaska, to Leningrad over the North Pole by dead reckoning on a special Alaska Airlines flight. Upon landing in Leningrad he thought he was back in Edmonton. The terrain and the people were very similar in the two cities. He set about writing comparisons—and discovered why Canada hadn't yet adopted Communism.

U.S.S.R. – *The Calgary Herald* – August 7, 1971

To effect a rise in the living standards of 240 million Russians, the Communist Party of the Soviet Union has brought in a beefed-up agricultural program. Alexei Kosygin, chairman of the Council of Ministers of the U.S.S.R., directed the production be increased 20%. Power is going to figure more in agricultural areas. The amount of electricity available to agriculture will be 75 billion kilowatt hours. This is a 50% increase. The number of kilowatts generated for the country as a whole will be doubled and large plants built in Siberia will join the power grid.

As I traveled throughout the rural areas during my two-week holiday in Russia, I could see many new power lines lacing the rural areas. It appears the Russians may follow the lead of agriculture in other countries and go into an expanded program of mechanization. The tractor fleet will be increased by 540,000 units—or a 27% increase. Big new tractors will be delivered from Volvograd, Kharkov, Minsk, Chelyabinsk, Altai and Leningrad plants. The machinery received by collective and state farms is put into a pool and drawn upon when needed. Anyone who handles machinery knows pools are the greatest destroyers of good machines ever devised. The Russian farm machinery pools are no exception. Kosygin gave notice that the machines must be kept in better shape. There is no use sending out new tractors if they are going to be "down" a large percentage of the time.

Another thing that disturbed him was loss of product after it is harvested. He said there must be a sharp reduction in these losses at all stages of production, storage and transportation. Plans are thus underway to enlarge the capacity of storehouses and the amount of refrigeration space. "An end must be put to the situation in which labour has been expended on product and it does not reach the people," he told the 24th party congress. Another major task with which he taxed farmers was to secure "full introduction" of crop rotation and to sow only high-grade seed. He said a tremendously important role in boosting the production of crop and livestock farming belongs to selection. "The country expects our breeders to evolve high-yielding plant varieties as well as the most productive breeds of livestock," Kosygin said.

In 1975 agriculture will get 75 million tons of mineral fertilizers and feed phosphates as against 46 million in 1970. He intends to see expansion of soil testing. By 1975 the output of mixed feeds will be increased to 30 million tons. Another thing that is about to happen is amalgamation of collective farms to effect better management practices. He talked about instituting confinement-rearing practices for more efficient production.

No doubt North American farmers will smile about the management practices that are to be introduced. On the few trips to the rural areas I made, I noticed there was a great deal of hand labor and relatively little mechanization. The only confinement rearing I observed was near Sochi, where a man was holding a sheep on a string and cows and calves were tethered along the roadside. In this area skinny spotted pigs, which looked like Landrace, were turned loose in spring to wander in the bush all summer. They were expected to return with a litter in the fall. Their main food was acorns. Turkeys were wandering all over the place, too. In this area also were thousands of hives of bees tended by apiarists from the state farms. The hives were like a small casket with a wooden carrying handle for moving them from place to place. The tenders lived in tents among the hives.

All this doesn't mean Russian agriculture is not successful. That country's agriculture suffered terribly during the war. It still regards itself as a developing nation. It admits it has a long way to go to catch up to nations who have industrialized their agriculture. Its propagandists are quick to point out Soviet agriculture wouldn't have come back as fast as it has except under a Socialist system. Only last year did agricultural production reach the pre-war level.

Research in agriculture is not being neglected. In Samarkand, the ancient cradle of civilization, there is located the only Karakul sheep research institute in the world. A researcher here perfected a hormone injection two years ago to make the ewes produce two lamb crops a year. Incidentally, the Karakul will walk 40 to 50 miles a day looking for water.

In 1966 my friend, Don Baron of The Country Guide, Winnipeg, spent three weeks looking at Russian agriculture. His conclusion was that the only picture one could bring home is one of a serious and determined people moving ahead rapidly to build a more productive and stronger agriculture. From this conclusion he foresaw "shadows of fearful competition" facing other nations producing agricultural products. The figures from the last five-year plan looked impressive to him. The figures from the new five-year plan look more impressive still. If Russian agriculture does surge ahead it will be only if farmers are given more incentives than these. The Russians, it seems, don't have an active export policy. Their main concern, according to Kosygin, is to get more food to their own people. This won't happen in this five-year plan or the next. The proper incentives don't seem to be there although the official attitude toward agriculture is good.

U.S.S.R. - *The Calgary Herald* - August 10, 1971

I had always imagined a Russian collective farm was like an oversized Hutterite colony in Alberta and that a state farm was run on a basis similar to a Prairie Farm Rehabilitation Administration (PFRA) community pasture. A number of basic differences were evident during hurried visits two weeks ago to the Red Uzbekistan Collective Farm on the outskirts of Tashkent and a tea-growing state farm near Sochi. The visit to the collective farm was unscheduled. It came about as the result of a visit to the city of Irkutsk in Siberia being cancelled because of flooding conditions. We were held in Tashkent and told that we'd be taken to this collective farm for lunch. I had visions of going into a Hutterite-type dining room and being able to meet some farmers. But this was not the case.

We left the city, went past a big old estate owned by the government for the entertainment of visiting firemen like Prime Minister and Mrs. Trudeau and continued along a tree-lined paved road through a suburban type community with stores and homes. Here and there were factories and there was an excellent public transportation system. In about half an hour we arrived at what appeared to be a suburban shopping centre (only the stores were smaller). The buses drew up before a two-storey building, which I took to be a golf club or country club.

"You have been driving through the Red Uzbekistan Collective Farm since we left the city limits," said the local "Intourist" guide, Lily. She was a young, smiling, informative Eurasian woman with over-size blue sunglasses and had been assigned to the press corps during the recent Trudeau state visit. "This is the farm's rest home. It is used for entertaining visitors, social events, wedding receptions and recreation." The only farmers in sight were few people eating large bowls of soup with black bread in the ground-floor restaurant. Our group and another touring group from the southern United States partook of a leisurely lunch at which a group of Uzbek natives were called in to play native music, sing, and dance for us. It was somewhat like calling in the local Blackfeet to do a chicken dance for foreign tourists at the Banff Springs. However, I understood these "natives" had a steady job of entertaining professionally every day at noon.

Later we made a rather hurried bus tour of the farm along paved roads but it was hot and the city-oriented group wasn't much interested in the farm operation. The surprising thing about this farm was that although it has only about 8,500 acres, some 8,000 people lived and worked on it. I suggested some of the workers must have been going to work off the farm in town but I was assured this wasn't the case. This collective farm was described as one of the richest of its kind in the U.S.S.R. It was no doubt a showcase for the benefit of tourists. Its chief production was fruit and vegetable crops on irrigated land. But even so those 8,000 people on only 8,500 acres of land must have been stumbling over each other most of the time.

Artels have been part of Russian life since the 19th century. Artels are associations of independent workers for collective work with a division of profits. I was always under the impression that the revolution was based on the theory workers could force into agriculture by the simple expedient of nationalizing farms, and applying labor union principles. I had a further impression the collective was obligated to look after the collective farmer from the cradle to the grave and, as a means of repaying the state for providing for his needs, he was supposed to produce to the limit of his ability without financial remuneration. It doesn't seem to have worked out quite like that.

Under successive five year plans the collective farms have been told to produce more than the ability of the farmers. That wouldn't have mattered much had they been able to operate on the community of the goods principle. However, the Communists under Stalin invented a nasty bit of business called piecework, merit award and incentive pay and a few things like that, which American unions have tossed out of their contracts. The average worker (man or woman) on the collective farm today earns about $120 a month and has been promised higher wages, pensions, and a few other welfare benefits, and the chance to sell more produce off his own one-acre plot. It's hard to determine if anything has gone sour with the collective or kolkhoz system of farming in Russia today. However, the peasants, who believed in the Leninist way of life and have had a chance to sample it on the farms for a couple of generations, seem to feel there are other more important things to do than raise food. Many young people have no inclination to stay on the farm.

National leaders like Kosygin and Polyansky, it seems to me, have been forced to admit today's farmer is not the same peasant Lenin was dealing with in 1917. The only close parallel is the Western Canadian grain farmer. The older growers still remember the struggle for equality of markets and pricing. They fought for change. To them, the beloved Canadian Wheat Board system of selling, which replaced older methods, is "the greatest." Events of the last few years have seen a new generation spring up who believe some alternative marketing and pricing methods look "great," too.

Somehow the Hutterite system has created an incentive for expansion. The Russian government has not discovered this secret. In each successive five-year plan the kolkhozs are urged to expand production to greater targets. They don't do it automatically. In fact, having exhausted some incentives, the government is now appealing to the patriotism of the people to maintain the kolkhozs. What the great experiment in Leninist Socialism has shown—and this has been revealed in speeches of today's Russian leaders—is the peasantry really didn't want to assume the responsibility of being land owners. They were promised something for nothing and they're still looking for it.

Just as our Socialist-oriented Alberta politicians have been successful in convincing people they can give them "free" this and "free" that at election time, so did several of our "Intourist" guides try to impress upon us the same tired old story. The greatest benefit of the new five-year plan was "free" schooling we were told in Leningrad. Another benefit was that those earning less than $75 month were not now required to pay income tax. "Tell me about this income tax. Does it pay for free schools?" I asked in all innocence.

"No, they are paid from the national income," replied our well-schooled guide in all seriousness. The national income is derived from the state buying farm commodities low and selling them high. The state is the only place farmers can sell their produce.

U.S.S.R. – *The Calgary Herald* – October 15, 1971

Following the inspection of the new Swift Canadian Company beef-killing plant at Lethbridge recently, a group of visiting Russian scientists specializing in food processing sat down with management to ask questions about plant operations. Their line of questioning was surprising for a group of persons from a country with a Socialist government and outlook. In fact, had one not been aware of their identity beforehand, he could have easily mistaken them for a group of free-enterprise businessmen from a capitalist country. Their main line of questioning concerned the economics of plant operation. What could be considered a normal payback period? When would the plant begin making a profit? What rate of depreciation was allowable for tax purposes? The questions on profitability and cost-accounting procedures came thick and fast.

Premier Alexei Kosygin made no bones about the fact industry in the Soviet Union was expected to become more efficient. Under the new five year plan he said economic stimulation of the country would be based on strengthening and developing the profit-and-loss principle. The government now regards profit and profitability as import-ant indicators of the effectiveness of production. Profit is the main source not only of funds of enterprises and amalgamations operating on the profit-and-loss basis but also the most important source of state budget revenue, Kosygin said.

Thus in line with this directive the scientists were busily investigating Canadian businesses, which depend upon profits to keep them going. Another new Soviet policy, I discovered while in Russia was confirmed by the mission at Lethbridge. This is that 30% of all state investment in the economy will go toward the development of agriculture in the next five years. As the result of this investment Russian farmers are expected to increase production by 20%. Figures like this make Canadian farmers drool. Our governments are trying to reduce their investment in agriculture. The Russians would increase theirs. It is a common Russian adage that howsoever high the spirit of man may soar, it is on the stomach that man, like the army, must advance. The ultimate aim of the new five-year plan is to increase all kinds of consumer goods. The Russians realize that since most, if not all, of the needs and wants of man are purchased with food, they must of necessity push food production. After 50 years of Communist rule, the Russian workers and leaders are beginning to realize their country has a terrific potential but that potential previously has been in the hands of incapable inefficient masters.

One of the Americans on my recent Russian tour made this observation: "It is a little thought-provoking and a little chilling for us to realize the waste that is going on in Russia and the superb efficiency of the inefficient!" This American characterizes Russia as being the largest and most potentially powerful nation on the globe. It is thus in this strange context that we find a group of their scientists visiting Canada under the guise of carrying out research of mutual benefit, but in reality attempting to assess the merits of capitalist methods of production. They confess to J.F. Gough, manager of the Lethbridge meat plant, their terms of reference are to double the amount of meat available to the Russian workers. They know our per capita consumption of meat is higher than theirs and they have been told to investigate whether it will be good for human nutrition to raise theirs and eat less bread.

This investigation is predicated on being able to produce and process more efficiently and being able to deliver larger quantities of consumer goods to the workers. They were anxious to know from their Lethbridge hosts what kind of incentives the workers were given. Are they all paid the same wage? What are their wages and can the high wages Canadians are paid in comparison to Russian workers be justified on the basis of productivity? They wondered if working in an abattoir was seasonal work. Apparently this is a problem in Russia because there are more grass-fed cattle there and more cattle are marketed at some seasons of the year than others.

They were interested to know what happened to sick animals; if it was necessary to pay the farmers for sick animals. (The sick animal here is the responsibility of the farmer as long as it is in his hands. If a livestock buyer makes the mistake of unwittingly buying a sick animal it is his mistake and the packer takes the loss. Some sick animals are bought subject to kill, i.e., if the meat is rejected on federal inspection, the farmer doesn't get paid.) The Russian scientists wanted to know how Canadian cattle were kept during the winter. They seemed unfamiliar with North American method of grain finishing, which is carried on all winter—outside in feedlots in the West and in barns in the East.

It was a little puzzling to them that Swift had no connection with Burns Foods Limited (they had inspected a Burns packing plant at Kitchener, Ontario, several days previously). To them the idea of competition was somewhat incongruous. Their theorists have impressed upon them for several generations now that competition is wasteful and that one chain of government-owned packing plants is all that is necessary. What they discovered, though, was that there are elements of efficiency in competition. They also discovered that competition provides certain elements of incentive that their Socialist society does not foster.

In fact, the Russians are finding out that it is impossible to stamp the competitive instinct out of man. They discovered that a collective farmer can sometimes produce more food on his one-acre plot than he will on 50 acres when his only incentive is working for the glory of the state. The competitive instinct to accumulate worldly goods just can't be stamped out of the human spirit and instinct. Having made this discovery, the Soviet leaders have evidently decided to expose their scientists and technologists to it in North America and then allow them to introduce it on a limited basis.

U.S.S.R. - *The Calgary Herald* - January 30, 1986

Some time next month, the 27th congress of the Communist party of the Soviet Union will lay down a new five-year plan designed to double the economic potential of the country and carry all areas of life to a new and higher plain by 2000. Some revolutionary programs will be introduced for agriculture by Mikhail Gorbachev, the Soviet leader. The party leadership wants to push agriculture out of the doldrums, especially in grain production.

"We don't intend to do it like the Chinese did by radical liberalism of rural policies, i.e., allow individual farmers to increasingly use capitalism," I was told by a member of the Soviet Embassy in Ottawa. From what I can gather, a new system will be tried out in the Krasnodar region. It will not reward the individual but the state or collective farms themselves. They will now be able to keep all the profits they make in order to expand and stimulate production. The collectives will also be given more autonomy from central planning to finance and run agriculture enterprises. This, it must be agreed, is extremely revolutionary. (It was revolutionary because all production hitherto had been sold to the state to, in turn, sell for funds on which to run the state.)

Writing in Soviet News and Views, published by the Soviet Embassy, Gennady Pisarevsky said so many people had taken advantage of the system that something had to be done about "parasitism, neglectfulness and irresponsibility, lack of discipline, unwillingness to do one's best on the job and even direct evasion of productive labor." Here in Canada, our governments have taken action against drunks getting behind the wheels of cars and killing people. It is hard for Canadians to visualize Soviet government action being necessary because farmers and farm workers are drunk so much of the time there isn't enough food. The government action put drinking out of fashion. In Canada, the farmers are the most sober and stable element of society. Politicians have probably secretly wished more of them would get drunk and wouldn't pile up so many embarrassing food surpluses for them to deal with!

The Soviet leaders have therefore decided to wean the people off the bottle by giving them something else to do. They will get a piece of the action by making money for the collective firms. They have discovered they don't really care about the state but they may care about the farms. They will use their competitive instincts to try to do better than the neighboring collective.

U.S.S.R. - *The Calgary Herald* - June 5, 1986

One of the most analytical dissertations on North America's weather problems ever given in Calgary, was presented to the annual meeting of the Western Barley Growers Association by Dr. Tim Ball, a climatologist at the University of Winnipeg

I wrote about this topic in the March 18 column. A few days later I received a letter from Les Rideout of Red Deer, founder and retired publisher of The *Ad-Viser,* a free distribution paper in the rural areas. (Rideout died in the Red Deer Hospital on May 27 at age 70.) Rideout found Ball's views interesting but sent along a publication called *Youth Action News* of Alexandria, Virginia. It is one of a number of extreme right-wing publications in the U.S. aimed at seeing that the Yanks don't go soft and fall for Marxist-Leninist communism.

The principal story was entitled: Soviet Electromagnetic War Actions. I read through the pamphlet and dutifully put it in file 12 1/2. This is the one where ideas for a rainy—or snowy—day are saved. On May 15, I pulled this pamphlet out of file 12 1/2 to see if it would reveal any new truths about the violent weather system that overtook southern Alberta starting the afternoon of May 13. This column was written May 15 when the ferocious storm left me isolated and without power for 36 hours; no telephone for 12 hours; no mail for days; a foot of snow and five-foot drifts and more wood to cut to stoke our free-standing fireplace than on any day during the winter.

There were no newspapers as country deliveries were suspended by snow-choked highways, and gasoline supplies were at a minimum due to electrically operated pumps being out of commission. The only thing working was natural gas under a cooking top. The lawn mower was buried under a four-foot drift after being used for grass-cutting the afternoon before the storm. There wasn't any way of getting in to Calgary to pick up the garden tractor at the repair shop. We had been using winter onions (shallots) and asparagus out of the garden the day before the storm. Now here we were, after 36 hours of continuously howling wind, completely cut off from everything.

The Youth Action News provided a plausible explanation of why the garden-planting weather had deserted us and brought snow and below-freezing temperatures. The story by C.B. Baker more or less followed the script used by Tim Ball, but it went one step farther and gave the reasons why the jet stream—the chief governing factor of weather hereabouts—had dipped so far south of its normal pattern May 13. The cockeyed weather patterns were the result of use by the Soviets of Nikola Tesla's theory of man-made earthquakes. Lieutenant-Colonel Thomas Bearden, a U.S. nuclear engineer and authority on Tesla technology, told the whole story to the 1981 conference of the U.S. Psychotronics Association—but nobody paid much attention.

Now that people are going back to see what Bearden had to say they are struck by the abnormal weather patterns and earthquakes since then. They theorize: the earthquakes have been triggered by extreme low frequency vibrations through the earth's core and "airquakes" have turned winter into summer and vice-versa. I am sure that if all the airborne energy, which practically knocked out southern Alberta for three days in May, had been generated underground it would have resulted in an earthquake of considerable proportions. The Soviet weather zap attacks, known as the Siberian Express, are expressly designed to disrupt North American agriculture either by drying it out or freezing it out and warming up the northern areas so that crops can be grown in the Soviets' northern regions. The Soviets have weather-blocking systems to interfere with the jet streams and produce large hot spots or large cold spots in given areas in North America.

All these outpourings of Bearden and *Youth Action News* may sound like tripe and crackpot nonsense and could raise doubts about a columnist who repeats them in an enlightened, estimable family newspaper. I'm not asking anybody to believe these ravings written in the middle of a May blizzard. Just believe there was a world-class blizzard. However, if anyone has a better theory on what happened the night of May 13, write it up and send it in.

Some of the countries Schmidt visited, such as Afghanistan, are no longer tourist destinations. Starvation is even more prevalent today than when he first saw its tragic face there in 1974. An agrologist friend in Drumheller, Alberta, gave him an introduction to an Australian working in the U.N. compound in Kabul. But, when he tried to get in to see the Australian friend in Kabul, an officious receptionist demanded: "Do you have an appointment?" Schmidt mumbled a few quick invectives under his breath as he made good his departure.

AFGHANISTAN – *The Calgary Herald* – August 19, 1974

On August 10, while on a vacation trip to Afghanistan, I stood on the walls of the ancient city of Balkh. This city is given credit as being the "mother of cities." It was here Aryan civilization began 3000 B.C. and developed into a cultural trade centre. After a similar visit to the giant walls, towers and mounds of mud brick in 1960, Arnold Toynbee, the British historian, said they gave him a "sense of the momentum of human effort on a grand scale sustained over a dozen centuries."

Within the walls of Balkh proper at its glory 12,000 persons lived. From here civilization started to go downhill toward the unmanageable and undefendable concentrations of population like Tokyo, New York and Sao Paulo. I didn't realize how totally defenseless the institution of the city had become until my memory took me back to the same date seven years ago when I stood on another of the world's historic sites in Northern Canada. The site was Port Radium in the Northwest Territories. Great Bear Lake is the spot from which man mined the uranium oxide that was used to build the first atomic bomb at the University of Chicago stadium. The bomb may have spelled the beginning of the end for cities.

On the YAK 40 jet flying into Balkh from Herat, a young American from Los Angeles, who was "traveling around" the country where the "mother of cities" is located, asked if I had heard THE news. I replied in the negative, not having seen an English newspaper since Teheran four days previously.

"NIXON HAS RESIGNED," he said triumphantly.

"When?" I asked

"The night before last," he replied.

"Will the world go on?" I asked.

He looked puzzled.

The resignation of the president of the United States seemed so irrelevant when put in the context that it intruded in a historic moment in this nomadic writer's life. Very few are given the opportunity to stand in the world's two most historic spots in the space of a generation. I didn't try to explain my feeling of humility.

Having published three non-fiction books, Schmidt is currently in the process of unleashing his considerable talents to write a fictionalized account of the military activities that occurred on Canadian soil during World War II. He spent twenty years researching and interviewing those who, in real life, participated in the Canadian North on the Canol Project, to reveal little known facts about Canada's role in the Manhattan Project. Schmidt holds the optimistic view that the atomic bomb has kept the world's cities intact from the time that humanity first witnessed its awesome power for destruction.

EUROPE

While writing about the agricultural issues in Europe, Schmidt can trace some of his own personal heritage. Similar to many Canadians, his roots are a blend of various European nationalities. In Scotland, home of some of his ancestors, he found them fighting inflation. In Ireland, where he has other ancestors, he found the Irish vigorously engaged in fighting each other, as well as strenuously arguing to move more beef into Canada. When he arrived in France, the home of still other ancestors, he found the wheat growing area much larger than he expected. In fact, some of their wheat fields are as large as those on the Canadian Prairies. And, the first time Schmidt visited the land of his German ancestors, he saw farm compounds where the farmer walked into one door to eat and later, through another door of the same building, to feed the livestock. Canadian health units, please copy.

ENGLAND AND SCOTLAND
– *The Calgary Herald* – August 22, 1969

The problems faced by British farmers are not too dissimilar to those of Canadian farmers, I ascertained during the past two weeks of holidays in England and Scotland. Inflation isn't a problem confined to Canada, although it seems to be galloping faster in Canada. Inflation hurts the small producer because he has fewer resources upon which to fall back and because his income isn't large enough to buy the necessities life. Often he is producing commodities for which the market has dried up or disappeared.

In discussing this problem in *Farming News*, published in Perth, Scotland, Morris Pottinger points out the returns to marginal farmers in the past 10 years have declined relative to other farmers and the community at large. "And," he added bitterly, "if these farmers are to prosper again—as they did during hungrier wartime days—the central authority will have to institute some radical changes in their thinking." What made the pill even more bitter was Pottinger's realization that "home production is not regarded as really important now that our former enemies, the Germans and the Japanese, are lending us money from their well-filled coffers."

Pottinger says the problem of the marginal farmer is a social one and cannot be solved by the standard means of giving production subsidies. "I think in many cases," he said, "agriculture grants and subsidies have done more harm than good and, although given with the best of intentions, have actually put agriculture on the road to ruin. They must be replaced by some other system better designed to bring marginal farms back to a healthier state." Pottinger believes the policy makers will have to throw away the rulebook in dealing with the marginal farm situation. As he sees it, Great Britain is going to need those marginal farms and the people who operate them in one form or another, sooner or later.

Ironically, although some of the "lower forms" of farmers appear to be in surplus position, there is a generally expressed fear that so many farmers have left the farms the U.K. is rapidly approaching the point where annual production is going to fall rapidly. Decrease in annual output is not an issue which Canadian farmers have to face—but lack of skilled technical farm help is. Good farmers simply don't like to see essential jobs on their farms going undone for lack of competent help. What benefit is there in gaining higher returns per acre while allowing essential work to be neglected? It is a bit disturbing for some of the older farmers that they cannot now afford to keep their farms in as neat a shape as their fathers and grandfathers.

Sixty-seven percent of all viable farms in the U.K. don't have any help and I imagine the figure in Canada is much higher. Alberta, for one province in Canada, has made a small start at a farm technologist training program at its agricultural and vocational colleges. Some of these programs were undertaken at the behest of farm commodity groups and have the inactive support of the Canada manpower department. In Great Britain, farmers have had forced on them (or have accepted) an annual $8 a year levy to support a government training program. However, it has become bogged down in quarrels over whether the program should be run by the agriculture department or the manpower department. Alberta farmers are fortunate in having their farm technician programs operated by the agriculture department in schools under department jurisdiction. These kinds of training programs don't come as easily in other jurisdictions where farmers are forced more and more to use the bludgeon to gain recognition of their needs.

A British Broadcasting Corporation television program sized up the farm labor situation as one where agriculture will have to start from scratch again and operate a long-term recruitment policy. "Farm leaders and farmers themselves," said the commentator, "must go out and try to present agriculture as a worthwhile career—and this means convincing government employment officials it has something to offer to the school-leavers. In terms of job satisfaction it can certainly outbid an eight-to-five factory job for a start. "The question of status is very much more difficult and subtle since, at least for the present, the farm worker is made to feel he holds down an inferior sort of job—a last resort job. This stems from the whole official attitude towards farming in this country—the atmosphere is at worst hostile, at best indifferent. Until farming is seen as vital to the nation's economy as, say, car production, we cannot look for a fundamental change here."

Seems to me this has a familiar ring about it!

IRELAND - *The Calgary Herald* - February 27, 1985

One of the most ludicrous commentaries I ever heard coming out of a radio was by a woman with a bleeding heart from London. She was trying to pin blame on the Mulroney government for unemployment in Ireland. It seemed that Irish packing plant workers were putting in overtime on Sundays slaughtering cattle and shipping beef to Canada. That nasty Mulroney, he with a good Irish name, had made a law clamping down on Irish beef imports. The law had thrown these poor, bedevilled killing plant workers out of work. And what do you think of a prime minister named Mulroney for cutting his cousins' pay packets?

No mention was made in this sad tale of despair about the cattlemen of Canada—and particularly those in Alberta—who think Mulroney is a prince of a fellow. He is simply doing what they have asked him: enforcing quotas to keep Irish beef from making Canada's hard-pressed cattlemen into non-taxpayers due to having their price unfairly cut. Canada allowed an annual quota of 145.1 million pounds of boneless beef into the country. That is a fair shake for all the exporting countries. Ireland was drifting along shipping an average of 5.9 million pounds of beef into Canada annually over the last five years. I understand the cattle which supply this beef are raised on shamrocks and the beef has a delicate flavor, but Canadians never get to taste that flavor as most of their beef is shoved into the hamburger grinder when it gets here.

Ireland's product is highly subsidized by European Economic Community moneybags in Brussels. This is the unfair part of these exports. Good-natured Canadian cattlemen had no complaints about this small amount of Irish beef arriving in Canada, but last year when the subsidized production jumped 47 million pounds, they started pawing the ground like mad bulls. The thing that caused their anger was that whereas Canada had set the import quota at 145.1 million pounds of beef, a total of 163.3 million came in—18 million pounds over-quota. Ireland was the largest over-quota offender. In its generosity, Canada raised its quota to 146.6 million pounds for this year but held Ireland back to 5.9 million. Importers, mostly Canadian packers and wholesalers, now require import permits from the federal government.

FRANCE – *The Calgary Herald* – August 17, 1982

The thousands of peasant farmers of Europe have enough political clout that they can write their own ticket for subsidies. As was explained previously, the farmers of the European Economic Community (EEC) manage to extract about $12 billion a year from the European Commission in Brussels—and they're always asking for more. This year the French farmers want a 16% increase in farm prices from EEC and other concessions from Agriculture Minister Mme. Cresson of France. When they thought the European Commission was stalling, the French Farmers Union didn't send a delegation to Paris, 100,000 of them drove Audis and Citroens and rode buses there in a mass demonstration. It was the largest ever staged by French farmers.

A British farm writer who rode one of the buses said none of the farmers on his bus had ever been on a real live "demo" before. Neither had he. "We were probably a bit nervous and that might have explained the appearance of small bottles of cognac and calvados passing around the overcrowded bus," he said. "The Place de la Nation in Paris was already half full when we arrived. But it was the noise and not the crowd which struck me at first." The demonstration was well planned with different segments of farmers falling into place in the procession. They carried banners which made rude remarks about Mme. Cresson's name (it means watercress) and others which made it clear that the British should get out of the EEC and stop holding up price negotiations in Brussels.

Oliver Walston, the British farm writer, said he didn't know whether the five-hour procession frightened Mme. Cresson and/or the European Commission. He was disappointed that she didn't appear on a balcony to say: "If they haven't any bread, let them eat manioc." (In staying out of sight, her discretion was probably the result of observing the treatment of Agriculture Minister Eugene Whelan of Canada, who became the target of milk and manure hurled by Quebec farmers a few years ago in a similar demo in Ottawa.)

The French demo was led by tractors and accompanied by a cacophony of explosions and empty milk cans pulled along the cobblestones. At the old abattoirs at the Porte de Pantin, the crowd was addressed by the French Farmers Union President.

Walston came to this conclusion: "At the end of the day I was left with the impression that the people who really invented the modern revolution had not lost their talent for mass demonstrations. There had been no violence and not a single policeman was to be seen. My fears had been unnecessary. I have developed a taste for big demonstrations now and my NFU branch meetings back in Britain will never be the same again."

GERMANY AND SWITZERLAND
– *The Calgary Herald* – September 18, 1984

When I was a kid growing up in a small Ontario village we lived on the perimeter next to a small mixed farmer who worked less than 40 hectares. Needless to say we learned a great deal from Alex McLeod about how to milk dairy cows, slop pigs, roll cigarettes from "makings," drive horses and feed the hounds. Our house was less than 25 metres from his cow stable and, yes, it used to stink periodically—and there were flies. For a period, I had a couple of milk cows and chickens of my own in a barn about eight metres from our back kitchen door. Other farms harboring livestock and poultry were also near the townspeople. The farmers never lacked for kids to come around and help with the chores or drive horses. Living costs were next to nothing.

All that is gone now. Bylaws all over Canada prohibit keeping livestock and poultry within municipal boundaries. Even pigeons are banned. In that do-it-yourself age, before boards of health were ever heard of, people weren't so ignorant about the source of their food. Farmers were integrated into the community

Experiences from this kind of semi-bucolic upbringing came flooding back to me a couple of weeks ago when I visited Germany and Switzerland with a group of Canadian farmers. In those countries the small farmers haven't yet been moved out of the villages. That's because the barn is closer to the house than eight metres. The barn is part of the house. The farmer lives in one end; the livestock in the other.

Walking "downtown" in Meiringen, Switzerland one night, we came upon a rather long, well-landscaped house. At one end a door was open through which could be seen cow stalls and milking utensils and cream separator parts hanging on the wall ready for morning milking. Along the narrow walled streets of German towns, if you take a peek through the wooden gates you will often see a farming compound. The house is an integral part of the barn, or vice versa. They will tell you proudly this or that farm has been in a family 13 generations. Health units don't impede agriculture in Germany!

We had a visit scheduled to the farm of Paul Erich Etzel at Wehrheim, a "village" of 10,000. We drove through the centre of town, then up a hill into the residential district and stopped in front of a big double wooden gate. The gate was thrown open and there we were in his barnyard. The operation is 47 hectares on which he grows six hectares of corn, eight each of winter wheat, barley, oats, rye; the rest is grass for his 21 red Holstein milk cows. He fattens bull calves for veal, sells some breeding stock (some of his sires originated in Ontario out of United Breeders at Guelph) and sells milk to the nearby U.S. Army base.

During the summer the dairy herd is out on pasture in a field about eight blocks from the barn. The cows are milked with a portable milking machine. The calves are kept back in the barn. The barn, house and machinery shed are in a compound in which Etzel only has to walk eight metres from the front door to the barn. The neighbours' houses are up against the wall. We saw a dustman sweeping the chimney at a home on the other side of the wall. The mailman walked in off the street at noon. This establishment has been there for 400 years and will likely stay another 400 under existing laws.

The neighbours don't argue about what odours there are. However, they have been getting after Etzel about better maintenance of the old village wall, which surrounds Wehrheim and all municipalities. But he got in their good books by completely restoring the farmhouse to look like it would have in the middle ages. They live on tradition. There used to be 80 such farms in Wehrheim, but now there are only 20. It is common to see tractors hauling farm implements buzzing around the city centre streets going to the fields on the town perimeter.

The German government has a program of subsidizing farmers like Etzel to move to one of his fields and build a house and barn. But he won't. "This set-up works for me," he said. In Germany, where farmland sells by the square yard, he is a powerful and rich man, on paper. He could sell off the land for $1.5 million—but would likely have to sell it piecemeal to get the price. But he wants to pass it to his eldest son. However, at 45, he is burning out fast from overwork and wants to retire. He thus has a problem. When a son takes over a farm, he can't tell the parents to move to town. He must provide a place for them to live out their days.

Paul Etzel started out his farming days as a hired man hoeing sugar beets and managed to obtain a smallholding. The present farm had been in the hands of another family four centuries but the two sons in line of succession had been killed during the war. The place was rented until Etzel married the granddaughter, sold his small farm and took over.

Germany has larger farms than most countries in Europe, due to an edict by one, Adolf Hitler. The land tenure system saw division between the sons but this was reducing farm size to inefficiency. The Hitler decree was that land can be passed on only to the eldest son without division. But there are many two-cow farmers left.

The Etzel family figured I was a queer goose having a German name and being unable to speak the language. They understood how it was, however, after learning my family has been carrying Canadian passports for five generations. I am only the second one in the family to set foot on German soil since a widow brought her three sons out of the Kaiser's army some time in the 1850s.

EUROPEAN FARMERS
– *The Calgary Herald* – October 11, 1984

One of the reasons Eugene Whelan didn't get elected Prime Minister of Canada by the Liberal party (when it had the chance) was his espousal of supply management for agricultural commodities. Whelan not only had tobacco, milk, poultry products and wheat under supply management boards but he exported the idea to the European Economic Community for its dairy industry. That export took place last April and was quite a feather in Whelan's hat—but I never heard a word out of him about it during his election campaign or otherwise. It seemed odd he would snuff out word of his "victory." In fact, I did not become aware of it until I went overseas last month with a group of

Canadian farmers.

S.J. Fraser of Edgerton, Ashford, Kent, England, was the first farmer we visited. He runs a 150-cow milking herd on 100 hectares and, up to April, was doing quite well. Then the EEC, after inspecting the supply management system for milk in place in Canada, decided to adopt it—and put into effect a 6 % cut in quotas for the 10 member countries based on 1981 production. This was supposed to cut production to 99.5 million tonnes. There was a need to do something about cutting production of milk and other agricultural products in the EEC because the taxpayers were complaining about the $16 billion in subsidies being paid to farmers. In fact, if the subsidy payments had continued fears were expressed it would break the EEC.

Another EEC dairy farmer who suffered a quota cut was Paul Erich Etzel. We visited him in Wehrheim, Germany. His dairy herd is only 21 cows and he won't suffer any loss of cash flow because he runs his farm well and efficiently and has been set up in business for many years. He is making preparations to switch emphasis to a dual-purpose type of Holstein so that he can sell meat if they don't want his milk. Etzel is regarded as one of the larger farmers in Germany. There are many two-cow dairy farmers in that country who sell milk as a "cash crop"—and they will be badly hurt by this quota cut. His only complaint was the EEC quota cut came five years too late—after the oversupply problem had become overwhelming. He agrees with the EEC system—and agrees with a friend's assessment of it in the words of the poet, Goethe: "Where there's light there's also shadow."

If the Canadian supply management is anything to go by, the Europeans can look for a massive reduction in the number of dairy farmers. Just after the war, Canada had 400,000. Now it has 50,000. That's the "shadow" the Europeans have yet to learn about.

NEAR EAST

A chance conversation with an Israeli in Winnipeg and an invitation from the Israeli Embassy in Ottawa to host a farm writer from the Calgary Herald to go to Israel to study advanced irrigation methods resulted in Schmidt visiting that country several times. He found irrigation practices used there could help irrigation farmers in Alberta. Coincidentally, at about the same time, he also learned of an Egyptian delegation coming to Alberta to study irrigation projects here.

As a boy attending Sunday school in Ayr, Ontario, Schmidt used to give nickels and dimes to a local missionary, Tina Baxter, who worked in Indore, India. Visiting Indore years later he did not learn any more about the Bible, but he did discover a great deal more about acronyms for scientific organizations that expand food production world wide from Canadian agricultural scientists in Hyderabad. On a junket to Kuwait, Schmidt also learned a bit about the Islamic religion: Allah appears to look after drivers who never drive less than 85 miles per hour and pass other vehicles on curves.

ISRAEL – *The Calgary Herald* – January 17, 1973

One morning while at breakfast during the National Farmers Union convention in Winnipeg all the delegates gathered around wearing badges imprinted with Boycott Kraft. At the next table sat a portly gentleman who later identified himself as Harry Horlick, a resident of Eilat, Israel, literally one of the hottest spots on earth. He was naturally interested in the reason the farmers were boycotting Kraft Foods Limited of Montreal, as that international conglomerate no doubt trades in Israel.

It was explained the NFU had taken the action because Kraft had refused to bargain for price on manufactured milk with several of its Ontario locals. However, it was further explained, Kraft was not in a position to bargain directly because all milk in Ontario is priced by the producer-controlled Ontario Milk Marketing Board, and the board's actions are supervised and may be varied by the Ontario Milk Commission, a government body.

"This puts a different complexion on the boycott," Mr. Horlick said. "The boycott will never work. That is based on the premise you can't beat the state." He then proceeded to relate how Israel, as Socialist a country as is to be found in the world today, broke a milk producers' strike. The producers there are required to sell to a government-controlled marketing board. They wanted more money. They withheld product from the market. Using the state-controlled El Al airline, the government ordered an airlift of milk from several North African counties. In a week the strike was broken. "You can't beat the state," he reiterated. This incident took place during November.

Mr. Horlick had several other interesting observations about agriculture and Israeli agriculture in particular. Located on the Gulf of Aqaba, Eilat's temperature runs to 130 degrees. Residents learn to take a salt pill once every hour to keep their body tissues from desiccating. Under irrigation this area produces multiple crops of fruit and vegetables. Whereas in Canada cold soil is a limitation on agricultural productiveness, the problem in this area is hot soil. "In this warm soil radishes grow as big as apples in a short time," he said. "But they are coarse and pithy. Nobody wants to eat them like that. "So we are experimenting with refrigeration units to cool down the soil and reduce the size of the radishes and other vegetables. We also use this cooling technology in marketing: to regulate the growth so that crops will not all come onto the market at the same time and lower the pricing structure. The swimming pools have to be cooled down to 100 degrees, too.

He is involved in a project for supplying tomatoes to the tourist hotels in the city. The small acreage he has been assigned produces 2 1/2 crops a year. Mr. Horlick is a former Vancouver businessman who went to Israel to put his skills to work for the Jewish homeland. He is familiar with Alberta and Alberta pre-war politics. His present visit to Canada was involved with selling jam. The jam is made at a kibbutz where a family started making it in a small way and the enterprise has now expanded to international proportions.

I got the impression from him the kibbutz system was in a bit of trouble—and the trouble had mostly to do with affluence. It's a case of starting with nothing and working hard for six or seven years. At the end of that time there isn't enough work for everyone, the hours are cut, the management may begin to adopt luxurious trappings which cause worry to some of the members, and a sense of boredom or lack of incentive sets in. When this happens, the Hutterites in Canada channel their energies into forming new colonies and trying to buy more land to establish the new colonies.

Israel is brimful of brains and has many exciting agricultural and industrial projects under way, Horlick said. "The reason we are expanding in all directions is that university educated young Jews from other countries with no place to put their talents are coming to the homeland where their skills and technology are in great demand. They are given free rein to undertake projects they would be denied elsewhere so we benefit greatly from this importation of brains." They have some exciting projects under way, such as a successful plant to convert seawater to drinking water and tapping underground water supplies of brackish water for crop propagation in the Negev desert.

ISRAEL - *The Calgary Herald* - February 4, 1975

In Canada few persons of Jewish extraction are farmers. They are generally found in the agribusiness segment of agriculture, if at all. It is therefore surprising to hear in my travels through the agricultural community indications that Israel is where some of the most exciting things are happening in agriculture both on the farms and in research labs. This is happening in a country which is usually thought of by Canadians as a hot, rocky land which supports little vegetation. However, the best Jewish brains and workers from throughout the world have been flowing into the re-established state of Israel for 27 years. They have done some startling things in the wilderness.

A farm group on a study-tour organized by a long-time farm-writing acquaintance, Clare Burt of Brampton, Ontario, now turned traveler and travel agent, explained why the Israelis want to bother with such uninviting terrain. Part of the reason is historical, but the most pressing reason is the challenge. Israel wanted to lead the world in arid zone research, to start a farming revolution which will help to feed people in the future. The desert lands just need water and some fertility to commence producing crops. They are growing plants by the "drip" method of underground irrigation. Each cultivated plant receives only the required amount of water for optimum growth, with no water being wasted through evaporation.

They have thus extended the agricultural area deep into the hitherto desert waste-lands of the Negev. They have done this at a time when there is great speculation about the deserts of the Sahel region of Africa advancing south. Perhaps this technology will be exported there to reduce the fears of the inhabitants who have suffered privation and starvation in the last few years. The Israelis told the Ontario group they are utilizing all the available irrigation water from the Jordan River, bringing it from the Sea of Galilee in a 108-inch pipeline right down into the upper end of the Negev Desert. In addition, this water supply is being supplemented by strategically located drilled wells all over Israel's farming areas. Israel's desert agricultural production is also increased by developing new strains of traditionally "dry-land" plants, for example, wheat, which will produce satisfactory results in semi-arid areas. "We were told on good authority that the Israelis have become one of the leaders in this desert-type of farming. They have an arid-zone agriculture research station going full tilt and coming up with results," Burt said. "We were amazed to see thriving reforested areas on rocky hillsides and in more level fertile land market gardens, which for generations were only marginal sheep pastures."

The swamp areas were hitherto avoided by the Arabs for centuries because of malaria. Since the early days of Zionism when Jewish people came back to live in Israel they have developed techniques to utilize swampy land. In a systematic, methodical manner (not without some mishaps, of course) the Israelis now have this whole area drained and producing fantastic crops of all kinds.

Some of these farming areas would be worthwhile looking into by farmers with itchy feet looking for somewhere to travel. As a contrast to the dynamic agriculture forming the desert, in the Israeli ministry of agriculture building in Jerusalem is an imaginative agricultural museum. Set up by Moshe Dayan, when he was minister of agriculture in 1961, the museum depicts how agriculture was carried on in biblical times. It gives an idea how far man has advanced from his nomadic wanderings, just merely to gain enough food to survive, to the kind of agriculture needed today to support man in his jet-age nomadic wanderings. Incidentally, the museum is located in the courtyard of the mansion built in 1860 to accommodate members of Russian nobility during their pilgrimages to the Holy City. They built a really luxurious stopping place.

KUWAIT – *The Calgary Herald* – August 21, 1974

Whatever happened to Spiro Agnew, lately deposed vice-president of the United States? Everyone assumes he has hidden out somewhere to write a novel. Few know he is now engaged in what may prove a more useful pursuit for the benefit of mankind than his flyer at the vice-presidential post. During a stopover in Kuwait July 29 on my recent Middle East vacation, I learned Mr. Agnew was also in the city—with little fanfare. I doubt if it is possible to create fanfare anyway when the temperature is 120, as it was that day. Agnew represents a U.S. firm which is trying to work out a deal with Under-secretary Mahmoud Al Adasani of the Kuwait finance ministry and Chairman Abdul Baki Al Nuri of the Kuwait Petrochemical Company to set up a company to produce edible protein suitable for human beings as well as animals.

The talks are at an early stage and very few details have been let out. Nobody would say what his role in the negotiations is beyond the fact "it is a normal business proposition." It was Agnew's second visit to Kuwait since May. No doubt his principals can provide skilled manpower and technology for such a development. Kuwait has the dinars. This country along with Libya, Saudi Arabia, Abu Dhabi and Qatar, have a combined revenue projection this year of $40 billion. The Kuwait government can't figure out how to effectively spend its share of those oil revenues within its economy this year. It has 50 to 60% of its budget available for outside investment.

In some so-called underdeveloped countries, progress—or what Western civilization likes to call progress—has been slow because of a simple lack of energy on the part of the people. Lack of energy results chiefly from the lack of protein in the diet. Lack of protein is manifested by the swollen bellies one sees in certain countries. The simple explanation of this is that protein raises the osmotic pressure in the body and a lack of protein osmotic pressure in the blood means the tissues become "waterlogged." This, in turn, causes the swollen bellies. Lack of protein may also be responsible for a reduction in calories—and this has an effect on the brain.

The work Spiro Agnew is doing is for the benefit of mankind—no matter what his role. Thus it is not stretching it to say it is top-priority work. There are a number of influential people in the world who feel it is a crime to "waste" the remaining petroleum supplies on energy and heating houses. They contend other products can serve these needs as well. The Shah of Iran is one of these people. He points out that since there are 72,000 by-products of oil, its use should be confined to conversion to these by-products. Although Iran has some semblance of winter, a protracted spell of 30-below weather on the Canadian Prairies might well convince the shah there is some merit to using petroleum for heating.

The responsibility of investing the oil windfalls wisely and well is an awesome one for any government. Nothing like it has ever occurred with such dramatic suddenness in the world's history. Encountering Municipal Affairs Minister Dave Russell of Alberta on the return flight from a fact-finding trip on public housing and establishment of new towns in the Scandinavian countries, he admitted Alberta may experience the same problem. This province is dealing only in millions against the billions the Middle East countries have available. Already dozens of woolly-headed theorists have advanced their pet schemes for spending the extra revenues.

It's too early to determine whether hard-headed business sense will prevail. Some of the autocratically-run Arab countries have been criticized for the way in which they are putting their money to use. Some have long been recipients of foreign aid from Western countries. Now it's felt is the time for the Arab countries to pay some back into United Nations agencies, the World Bank and other relief funds. However, the sheiks and shahs are looking ahead to the day when oil supplies may be depleted and are investing their monies in "sure bets" in U.S. real estate such as resort properties, downtown business blocks and other real estate. The Shah of Iran, through his Pahlavi Foundation, has bought two buildings in downtown Manhattan for wrecking and constructing a 34-storey revenue building. He has purchased a block in Brisbane on which is located the Gresham Hotel (my home for a few days at one time) to redevelop as a revenue property.

There have been some pointed questions asked as to when America will be relieved of some of its foreign-aid burden. The Middle East leaders pointed out they are spending money on their own poor but they must also invest for their future. They are hedging their bets by not only buying into Western countries' real estate and industry but by buying technology and manpower from the West in food-producing ventures like that with which Mr. Agnew is associated.

Let's see now, where could the Alberta government invest a few million in Iran or Kuwait?

EGYPT – *The Calgary Herald* – October 16, 1983

The Egyptians, like the sphinx for which they are so famous, never learned anything about irrigation. When the first irrigation systems in Western Canada were started before the First World War, many of the engineers employed were brought in after having completed irrigation works in Egypt and other African and Asian countries. Stories are told about them wanting to use bricks in the construction as that was the material they used in Egypt where labour was dirt-cheap. They were finally persuaded to switch to reinforced concrete.

Now the Egyptians are here as part of a four-week technical mission to find out what they have done wrong and how they can once again make their irrigation systems work and produce more food. Members of the high-level delegation completed their tour in Calgary today after spending most of the week in and around Lethbridge irrigation districts. They spent previous days in Ontario and Quebec.

The tour was mounted by the Canadian International Development Agency (CIDA) and included a week of study in England. Assisting in arrangements was Irene Mathias, a CIDA project officer of the Egypt program and sister of Mrs. Linde Turner of Rosebud. The tour was the first phase of a proposed five-year integrated soil and water improvement project to be funded by CIDA. The mission familiarized the Egyptians with the resources and machinery available in Canada to help carry out the proposed project aimed at increasing productivity of farmland in the Nile Delta.

Egyptian agriculture is limited to the 2.8 million hectares of traditional farmland where irrigation from the Nile is feasible. Inadequate drainage and irrigation systems and inappropriate and harmful water management practices have fostered salinity, water-logging of the soil and reduced productivity. Alberta irrigation districts have experienced the same conditions but large amounts of know-how and money are being expended to correct the problems.

EGYPT - *The Calgary Herald* - November 24, 1983

For the sake of the world hungry, the United Nations should pass an order requiring every head of state to have a degree in agriculture. Perhaps Eugene Whelan, the new head of the World Food Council, could steer this through with the same agility as he showed getting Canagrex through. Or those mysterious fellows who are pushing for a new international economic order. For the last 60 years, national leaders without fail, from Joe Stalin up and down, have pulled government support out from under agriculture and put all their national treasuries behind industrialization. Of course, when they have nearly bankrupted the nations and have seen the error of their ways, they went back to agriculture.

Egypt is the latest country to join the parade of nations marching to Canada to find out how we produce food and do it well and cheaply. Egypt was the granary of the Roman Empire. The people ate well also during the days of the British Raj. When Gamal Abdul Nasser took over after independence, he became obsessed with industrialization, diverting investment from agriculture to costly and grandiose state industrial enterprises, most of which turned out to be disasters. The Nasser and Anwar Sadat industrialization programs resulted in increasing an already poor populace to a 2.8% growth rate per annum. And whereas Egypt was a food exporter up to 1970, she has become an importer today, a best customer for some agricultural exporting nations for wheat and beef. Had not President Hosni Mubarak put a stop to this trend, his country would have to import 80% of its food requirements in 40 years for 140 million people.

Some U.S. observers feel Agriculture Minister Youssef Wall of Egypt has swung too far in his demands on farmers now. He has announced a program to double corn production to 3.5 million tonnes, almost double rice to 2.33 million tonnes and boost wheat 80% to 1.9 million tonnes. Egyptian planners feel the peasants are overpaid and the government hasn't offered them enough for the production Wall wants. The Americans feel he won't get the crops. Nevertheless, the government has instituted a five-year plan to expand onto new lands and to increase productivity on the 2.8 million hectares of Nile delta land, which have become saline and waterlogged. It is doing this with assistance from international aid agencies like the Canadian International Development Agency.

A CIDA proposal would do an integrated soil and water improvement project on a 20,230-hectare site served by one irrigation canal. The plan would include structural and institutional changes as well as remedial drainage works. It would also mean substantial export sales for Canadian equipment and tile-manufacturing machinery and a full range of farm inputs.

The Egyptians sent a high-powered delegation to Canada for a month to study irrigation methods. The delegation discovered Alberta irrigation projects have experienced similar conditions to theirs, that is, water-logging by irrigation water rather than by subsurface water as in Ontario. They wanted to talk to irrigation farmers in the Taber-Lethbridge area where large amounts of know-how and money are being expended to correct the problem. A stop on this tour was the farm of Ted Allen near Taber. He is vice-president of United Grain Growers Ltd. and is a large grain and irrigated vegetable grower.

The Canadians found the specialists in the delegation to be quite knowledgeable. When they inspected Allen's onion field one of them said to him: "Your soil needs more potassium." When Allen assured him he had his soil tested and he had followed the recommendations of the experts, the Egyptian replied: "But you need more potassium for growing onions. They will develop darker skins, which will come off easier if you add more potassium."

INDIA – *The Calgary Herald* – September 1, 1977

Earlier in the summer I spent three weeks in India. I was fortunate in being able to talk to a number of Indian, Canadian and American agricultural scientists concerned with food production. I have always had a great deal of confidence in this cadre of scientists to keep food production ahead of population demands. That feeling was reinforced by seeing the work being done in India and by learning of the work of other international scientific groups.

In the past I have been critical of the gloom-and-doom utterances of some of the United Nations agencies on man's ability to feed rapidly increasing populations. The Food and Agriculture Organization of the U.N. has been crying wolf for years and is still doing so about the precariousness of the world food situation. But, while the U.N. agencies are making loud predictions that 1 1/2 billion people will be living on substandard diets by 1985, there are a number of other agencies quietly working away to search out long-neglected possibilities in agriculture in areas of the world in which it is hard to grow food, namely the band around the earth covered by the tropics.

It's hard to feed people in this band and make them as fat and sassy as North Americans. The climate makes food production and storage difficult, despite the fact people in the tropics don't consume as much food and don't eat as much meat as those in the temperate climates. I know my food consumption was cut in half during my stay in hot humid India. What really keeps food consumption at a lower level than the nutritionists would like to see is lack of purchasing power. The reasons for this lack are extremely complicated. The only way I can see of improving the situation is the infiltration of the political wing of the U.N. by jobless Social Crediters in Alberta. The Socreds' ideas about increasing the purchasing power for the masses has to be the answer the Third World countries are desperately seeking to bring the purchasing power of their people up to a humane level.

In the meantime, these countries in this globe-girdling tropical poverty belt will have to be satisfied with the likes of **CGIAR, ICRISAT, ICARDA, CIP, WARDA** and the like. These acronyms are not household words in Canada, to be sure, but they are important to the well-being of the tropics.

CGIAR stands for the **Consultative Group on International Agricultural Research.** It is the brainchild of the **Food and Agricultural Organization (FAO)**, headed by Edouard Saouma; the **World Bank**, headed by Robert McNamara; the **United Nations Development Program (UNDP)**, headed by Bradford Morse, and 32 governments and private foundations. Its terms of reference are to finance the resources of modern biological and socio-economic research for tropical countries.

With the formation of the **International Centres of Agricultural Research in the Dry Areas (ICARDA)** last year, the number of centres and programs is now 11. With administrative offices in Cairo, it will study farming systems in both cold-winter and Mediterranean-type climates relating to sheep, durum wheat, barley, lentil and broad bean production. The earliest of the 11 is the **International Rice Research Institute (IRRI)**, established at Los Banos in the Philippines in 1960. Its main task is to develop new high yielding varieties.

The others are:

The International Maise and Wheat Improvement Centre (CIMMYT) at Los Banos, Mexico, is the best known of all as it is given credit for kicking off the Green Revolution under Dr. Norman Borlaug.

The International Centre for Tropical Agriculture (CIAT) at Palmira, Colombia, has the objective of improving agriculture in the humid lowland tropics, especially in the growing of cassava, field beans, beef and swine.

The International Institute of Tropical Agriculture (IITA) was founded in 1968 at Abadan, Nigeria, on somewhat the same pattern as CIAT, except that it is confined to Africa. One of the chief concerns is to develop more intensive cropping systems to replace shifting cultivation now typical of tropical Africa.

The International Potato Centre (CIP) was formed in 1971 at Lima, Peru, to develop new varieties and to expand potato production wherever they can be grown advantageously in developing countries.

The International Crops Research Institute for the Semi-Arid Tropics (ICRISAT) was established in Hyderabad, India, in 1972. Although it studies farming systems and water management systems to benefit farmers, it also seeks to develop superior varieties, of sorghum, pearl millet, chickpeas, pigeon peas and groundnuts.

The International Laboratory for Research on Animal Diseases (ILRAD) at Nairobi, Kenya is concentrating on two major cattle diseases, theilerosis and trypano-somiasis (sleeping sickness carried by the tsetse fly).

The International Livestock Centre for Africa (ILCA) at Addis Ababa, Ethiopia, is working to increase the output of animals, especially beef cattle, through improved systems of production and better systems of range management.

The West Africa Rice Development Association (WARDA) of Monrovia, Liberia, is a regional organization of the West African government devoted chiefly to applied research and aimed at making the region self-sufficient in rice.

The International Board for Plant Genetic Resources (IBPGR) is at Rome and seeks to encourage and help co-ordinate the collection and exchange of plant genetic materials of potential usefulness in crop development programs.

INDIA – *The Calgary Herald* - September 15, 1982

A great many Canadian farmers have found themselves locked out of the market-place because their production units are too small. Murray Sharp, a hatching egg producer from Sarnia, recently complained to an Ontario farm writer that United Co-ops of Ontario hatchery division refused to renew contracts for eggs from his 1,000-bird flock because he was too small for the co-op to be bothered with. The co-op is concentrating its purchases of hatching eggs from large-scale producers (over 20,000 birds). It has left smaller producers such as Sharp nonplussed and bitter that their services are no longer required in the food chain. They have been desperately searching for an answer to that complex question.

A speech given September 1st by V. Kurien, chairman of the National Dairy Development Board of India, at the 10th international conference of the World Jersey Cattle Bureau in Edmonton shows that we may have to look to an ancient civilization for the answer. In 1946 milk producers in India's Kaira District went on strike against a private dairy monopoly on milk procurement. It cut the price to the extent that dairy producers couldn't make a living. The strike resulted in a dairy co-op established at Anand to which every producer could sell his milk. Milk is picked up from 325,000 producers in 940 villages twice a day for sale as fresh milk and processed into butter, cheese and skim milk powder.

In commenting on this development, Kurien said: "The problem involved in organizing purchases of milk from hundreds of thousands of producers, most of whom have only one or two litres daily, is a task which calls for a fairly demanding level of managerial and technical competence. Indeed, it is widely agreed the competence of Anand Co-op's professional staff—and, above all, their commitment to the members who employ them—has set an example for India's professionals not only in our rapidly growing dairy industry but also throughout agribusiness."

The co-op has undertaken to carry out sidelines such as the sale of trademarked milk, provision of mobile veterinary clinics, artificial insemination and breed improvement, seed, mixed feed and other inputs. The original Anand Co-op has now expanded into 59 co-ops in 16,000 villages with 2.2 million members since 1970. This only covers 15% of India. It is the aim of the dairy development board to set up 10 million more small dairy farmers with better livestock over the remaining part of India in the next eight to ten years.

The Indian dairy herd is like none other on earth. To begin with, it has been in existence for 5,000 years. Besides cows (most of them nondescript), water buffaloes are milked. The population of dairy animals is 87 million of which about 12 million are genetically improved cows. Much of the upgrading in progress is being done through the use of Jersey cows. Some 60% of the milk produced in India is buffalo milk. About 80 to 90% of the milk processed in the modern dairies for the marketplace is buffalo milk. The number of farmers is 90 million, a figure hard to imagine when compared to Canada's 318,000.

The upper class, comprising 30% of farmers owns an average of 13 acres (some of the real rich farmers have 25 acres). Another 20% are self-sufficient on an average of 3.6 acres. Another 30% are subsistence farmers with 2 1/2 acres and 20% are landless persons who depend on farm work to make a meagre living. But Kurein said the majority who produce milk have either no land or less than three or four acres. How does this important segment of 18 million farmers with no land run dairy herds? Kurien explained: Their method is quite simple. Their most valuable possessions are one or two Surti-type buffaloes, which are compactly built and docile. They thrive on coarse fodder and forages. The landless people who keep these animals earn a little money by hiring out as laborers for weeding fields and harvesting cereal crops. When they get employment they are paid partly in kind: they are permitted to take weeds and straw, which they feed to their buffaloes.

For much of the year, there is no gainful employment for any of the family members. One of the few productive activities for these unemployed or underemployed human beings is to garner whatever herbage they can find in the surrounding countryside for their buffaloes. They use this herbage in the hot dry months before the monsoons in July. Milk production has increased to 71 billion pounds in 1981 from 47 billion in 1970. (The 1981 figure is only four times that of Canada so dairying is a growth industry in India.) It is planned to increase the genetically improved milk cowherd to 25 million from 16 million in the 1990s.

In North America the political activists have been accusing us of squandering grain to produce meat and milk from cattle fed on it. They alleged the use of this grain for this purpose is robbing the hungry of the "most affected countries" of food. The experience of India knocks the stuffing out of their argument. We see about 30 million dairy animals fed on weeds and crop residues and leaves to produce protein-rich milk and meat as there is no taboo against slaughtering buffaloes. This material would go to waste were it not fed to the animals. Likewise, we have millions of acres of land in North America where cattle consume the same type of roughage and it would go to waste otherwise.

AFRICA

Happenstance brought Schmidt to southern Africa in 1973. Expecting to find vast areas of jungle, he was astonished to discover a huge and productive agricultural empire in both Zimbabwe and South Africa. Even Kenya is developing as an agricultural producer thanks to the United Nations agencies assisting in introducing simplified systems to their farmers.

African countries clearly show the necessity for an educational system that enables people to be trained for other kinds of work when subsistence farming is replaced by technology. The use of farm machinery greatly enhances food production, but without the skills to do other kinds of work forces thousands to languish in abject poverty without any means by which to sustain themselves.

ZIMBABWE – *The Calgary Herald* – August 22, 1973

What's the proper etiquette when a monkey walks up and bites your wife? Do you reach for a bottle of iodine or monkey-bite remedy? Do you bite the monkey back? Or do you invite the monkey for a spot of tea?

This was just one of the unusual incidents of our recent Rhodesia-South

Africa-Brazil holiday jaunt. It happened on Kandahar Island in the Zambezi River near Victoria Falls where we had gone on a game-spotting trip by riverboat with a group of sightseers. The tour guide guaranteed a troupe of 50 vervet monkeys would be on hand by producing a bag of peanuts. They came swinging out of the trees and followed the people around on a short walk then stayed for tea with us. Some of the bolder ones came down to seize biscuits and grab at handbags. Then for no reason, this one walked up and bit my wife on the leg and drew blood. There wasn't time to thumb through Emily Post. The guide seemed unconcerned. "Oh, you'll survive," she said. "I've been bitten by monkeys many times. Sit down and have some tea."

On that jaunt we looked in vain for crocodiles in that stretch of Zambezi. They were considered vermin because they had a sweet tooth for water skiers and had been killed off. So if anyone in Canada has a problem stretch of water due to water skiers, simply send to Rhodesia for a pair of big starving crocodiles.

The police use a fiendish but ingenious method to control illegal parking in Harare. They paste a large foot-square piece of paper over the windshield so the operator can't see to drive the vehicle way. On the paper in big red letters is a notice telling the driver he has broken the law. The driver has the devil's own time soaking the glue off the windshield. In fact, the only effective remedy is with a soft drink—once again proving the slogan that "every car goes better with Coke."

Walking on Copacabana Ave., I noticed corn falling from the sky. Looking up I observed a high-rise apartment and noticed every balcony had a pigeon cote. Later that day when eating at an outdoor cafe we were served some small mottled brown hard-boiled eggs as an aperitif. They were excellent and we concluded they came from the pigeons on the balcony. This might provide a specialty business for high-rise dwellers in Canada who think $1 a dozen is too high for chicken eggs.

Another food delicacy from both Rhodesia and South Africa is biltong. I'm not sure how this is made—whether it's air dried or smoke dried. However, it is strips of blackened beef that are chewy and sweet—and addictive like peanuts. Once again, this product might provide a specialty food business, which could catch on fast in Western Canada. I liked the comment of *The Rhodesian Herald* when all the soap-makers simultaneously announced increases in prices, an act that is happening regularly today. "We can understand a price war—but increasing prices out of sympathy for a competitor is nothing less than a peace offensive," said the editor, "and should result in an inquiry by the price-ring commission."

South Africa is the place to go into the broiler business. Sales are climbing at the rate of 20% a year. Col. Saunders of Kentucky hasn't even sent an expeditionary force to that country yet. Nine years ago the output was 10 million birds. Now it is 60 million. Per capita consumption has risen to 12 pounds in the same period from 8 1/2. The price has fallen to 25 cents a pound from 35. The main reason for growth is that Blacks are going for white chicken. Chicken is now the cheapest form meat available and when the Black industrial workers, arriving in the cities by the thousands, begin buying more fresh meat it is not surprising chicken is the favourite. The rising price of red meat has turned more people to the white meat of poultry. And another reason for increased chicken production is expansion and technical progress in the animal feed industry.

While Schmidt was on holiday, a friend, who took over his column for a week, wrote about the experiences he had while on assignment for CUSO in Malawi. The columns he wrote generated some controversy among a number of Canadian political figures. This column on Black African agriculture came from the pen of E.K. Nimitz. He was an Alberta farm writer for several years, being on the staff of Cattlemen magazine and Northern Thunder and writing at times under the pen name of the Ribstone Rustler.

SOUTHERN AFRICA - *The Calgary Herald* - April 15, 1978

Given the present food production levels and population growth situation that exists today, a country such as Canada—a net exporter of food in quantity—is a bit of a rare bird, indeed. Black African countries are not food-exporting countries for these reasons:

1. Population explosion, which overtaxes their annual agricultural harvest.
2. Climatic disasters like drought in the Sahel region.
3. Constant warfare being waged by liberation movements.

The man who has made the greatest mileage out of the word, "ujama," has to be Dr. Julius Nyerere, leader of Tanzania. He has attempted to instil the concept in his people beginning with the Arusha Declaration of 1967 that ujama is their natural heritage and the normal path to follow. Essentially ujama defines out in English as "African socialism" based on the extended family concept. Nyerere realized, of course, that the African is primarily a social animal and appealed successfully to Tanzanians in a charismatic fashion to accept the ujama concept as their own. Thus when ujama is discussed it is done so synonymously with Tanzania even though the concept is probably equally the heritage of many African tribes.

Villagization is the economic concept that Tanzania has tried to weld to the social concept of ujama to improve the lot of the smallholder farmer. This has been accomplished, Peking Chinese-style, by destroying the tribal structure and consolidating many small villages into one central village which practices communal cultivation. It works well as far as services go, but as far as farmers go many are not too happy to be walking five miles where they only had to go 100 yards before.

While the influence of the Dragon is being subtly injected in ways that improve the Tanzanian economy, like railways and agricultural advisers, the influence of the Bear is direct and obvious. Russia is taking the direct approach and getting involved and visibly supporting several groups of freedom fighters that are advancing the cause of the liberation movement. Three of the leaders of liberation movements have been dubbed the Red Crocodiles: Joshua Nkomo, Robert Mugabe of the Patriotic Front, and Sam Nujuoma of Swapo.

Canadians at this point may be unsure as to what the Africans are being liberated from, especially in light of the USSR's past record of liberating Jews and Ukrainians within their own country as well as other countries like Poland and the former Czechoslovakia. However, it appears there are some countries where Africans must be freed from the oppressive yoke of colonialism in order to be able to assume their proper liberated yoke.

While I was there, I noted certain organizations like CUSO, OXFAM and the World Council of Churches, encouraged by left-wing Canadian politicians, appear to have succumbed whole-heartedly and swallowed the United Nations line that freedom fighters and liberation movements are bona fide in nature. CUSO, for instance, has a pro-liberation-group attitude, investing in excess of $100,000 per year in these political organizations. Dare one suggest these well-meaning do-gooders climb down out of their ivory towers and reassess just exactly what the Red Crocodiles and their sponsor are really after? Do they want to establish some more unstable and chaotic economies to place a further burden on the world's food production system? Likely the folks in Angola and Mozambique are giving this a great deal of thought now.

Zambia is one of the better examples of a country in Africa that has concentrated on a single-commodity economy, copper in this case, and thus put itself at the mercy of wild price fluctuations on the world market. As the world price for copper is not all that hot, Zambia is, even after currency devaluations, still facing certain bankruptcy unless the World Bank bails them out with cheap loans.

Dr. Kenneth Kaunda or KK, as he is known, appears to have a basic weakness. He is a good talker and spends a lot of his time manufacturing rhetoric instead of getting on with the job of developing agriculture and putting his economy on a sounder footing. Even though Zambia has an excellent agricultural potential, KK and his humanism-oriented cabinet are still unable to produce any agricultural commodities for export that exceed the tonnage of rhetoric which is exported annually. Zambians have been described on occasion as a touch on the lazy side and this may have something to do with their current economic woes. However, KK does find financing somewhere to maintain ever-increasing armed forces. Another of Zambia's claims to fame is that several liberation movements find it a convenient place for their headquarters. This attitude and political activity has demoralized agriculture as far as the white farmer and the emerging African farmer is concerned.

Botswana is cattle country. About three million cattle are grazing on beautiful ranch lands similar both in appearance and productivity to the southwestern U.S. The main thrust of the development program is to develop the cattle industry with a series of smallholder ranches. Each ranch would run 500 head and water them from a central location. Approximately 5 to 10 African cattle producers will be involved in each ranch. Close ties are maintained with the Republic of South Africa because nearly all food, with the exception of beef, is imported from the republic. Botswana is also expanding its newly created defence force and relations with Rhodesia (Zimbabwe) are becoming tense, even though Rhodesian Railways provides the main trade system with South Africa. (End of guest column by Ribstone Rustler.)

KENYA – *The Calgary Herald* – January 16, 1980

In a previous column I suggested the Alberta government acted with undue haste in passing the Matrimonial Property Act, which came into force a year ago. Implicit in its haste was the desire to get pressure groups of screaming females off its back. I further suggested the MLAs didn't take the time nor were they given the opportunity to size up all sides of this socio-economic legislation which, in effect, forces farmers and small businessmen to take partners into their businesses the day they get married. If the marriage results in divorce, the business is to be equally divided. If such an agreement cannot be reached, a judge may decide on the division using 13 criteria to guide him in making the decision.

Other jurisdictions have considered similar legislation. Kenya is one of these. But it has gone a more thorough route than Alberta. The ruling Kenya African National Union party, headed by President Daniel arap Moi, has presented the Marriage Bill with 158 clauses to the Kenyan House of Commons. This bill is no hasty piece of work. It was presented in 1970 and was rejected. It was brought back last July and was discussed clause by clause.

Naturally, there was opposition from some of the male MPs. The opposition was summed up by Wafula Wabuge, MP for Kitale West, who said the bill would cause chaos and hardship if allowed to sail by unchallenged. He wondered why there is a pressing move to have some of the traditional laws rewritten. He alleged many clauses in the bill had no relevance at all to the African way of life. The bill provides for monogamous and polygamous marriages, the latter being traditional. This tradition provides that a man can marry 100 wives if he wishes so long as he informs the others. There is one exception to obtaining consent. That exception is he doesn't have to inform his first wife. This is where the male MPs ran into trouble with their female counterparts.

Dr. Julia Ojiambo, assistant minister for housing and social services, objected: "We do not oppose men to marry second wives. What we are asking from them is advance notice so we can be ready." Cornelius Myamboki, James Kimondo and James Kuria, parliamentary reporters for *The Nairobi Nation*, said the House broke out in cheers at this point; also when she added: "It is not strange that women ask advance notice as the African woman has for a long time respected the husband and will continue to do so as long as she is accorded this simple aspect of dialogue and care. It is not strange because most honorable members have more than one wife—and they have given their first wife notice of their intention."

The bill wants to get around the "embarrassment" of having the husband unilaterally "decide to marry another wife and bring her to the bedroom." Now, I ask you: Was there any cheering in the Alberta, Saskatchewan, Manitoba, and Ontario legislatures when their marriage bills were introduced? Did any woman MLA stand up and say Canadian society had long respected the husband? The MLAs were faced instead with loud cries about Canadian men being "male chauvinist pigs." Perhaps militant Canadian females have something to learn from their African counterparts. The Associated Country Women of the World went to Nairobi a couple of years ago for their annual meeting but none of its members brought home this story.

The Nairobi Nation's parliamentary reporters pointed out the bill "has been understood as a guideline for future generations where lifestyles will change due to industrialization." Wafula Wabuge noted that women tended to be possessive, mother-hen types. Slapping a wife is a "necessity," he said, "because by not doing so a wife will not realize she is loved." This was too much for Grace Onyanga, MP for Kisumu Town, who rose on a point of order and asked Wabuge whether he realized that slapping a woman could break her jaw.
But Wabuge would not back off. He gave the benefit of his experience: "If you do not slap a woman, you will note that her behaviour will not appeal to you. Just slap her and she will know you love her. This is when she will call you darling."

Laughter.

Dr. Ojiambo justified her stand on another clause in the marriage bill by soothing the male libido then angrily denouncing men for "spending all their salaries on drink and running a clandestine mistress downtown." She appealed to men "not to panic" when the bill provided that the woman should have the right to run the home through the credit system in the absence of her husband. She said the clause, providing that spouses had equal rights in sharing property, had been introduced at the right time: when both the man and woman contributed to creating their common wealth.

The view of Mrs. Eddah Gachukia, a nominated MP, was that the bill was aimed at strengthening the family unit. (Like Alberta, the Kenyan Parliament has only one party and persons who have been nominated to run against sitting members may show up and speak.) She said this would do a great service to the nation of Kenya. The real issue was not the kind of marriage but the requirement for stable marriages. The Kenyan constitution guarantees equality of men and women, but Mrs. Gacukia said the bill provided that men should assume their obligations, although Wabuge said the husband is responsible for all necessities of life and "therefore he is the boss."

The National Farmers Union in Canada, which was one of the lobby groups which pushed the matrimonial property bills in the provinces, also adopts the same position as the Kenyan women MPs. Mrs. Hilda Echlin of Holstein, Ontario, a member of the NFU women's advisory committee, said the NFU position is that legislation is aimed at strengthening marriage on the assumption that "marriage is an economic union to which husbands and wives contribute equally although their roles are different."

No new definition of marriage compounded by the National Farmers Union or its women's advisory committee members has ever grappled with the problem of how to divide the greatest asset of any family, the children. The best answer ever provided was by King Solomon—and he told the quarrelling women to cut the child in two. That's what is happening in a great many cases.

AFRICA – *The Calgary Herald* – March 7, 1986

Italy has taught the world lessons that could aid African countries in the elimination of that continent's food and economic crises. "In the 20th Century, Italy has proven that the poor agricultural regions located in the south could be successfully integrated with industry expansion in the north, thus bringing about orderly development," it is claimed by Edouard Saouma, director general of the Food and Agricultural Organization of the United Nations at Rome. "This is a masterly blueprint that could well be followed by others." He blamed much of the mess on the Europeans who colonized Africa. When they first went in they noted that food grew easily and life was easy. They assumed European-style agricultural technology could easily be transferred there. But much of the technology proved, inappropriate, mostly because of the superficial way it was introduced and because the human factor was forgotten. When production was started, agricultural and mineral resources were given away for practically nothing.

Winnie Ogana of Nairobi, a reporter with the *All-African Press-Service*, a lobby group, put it this way: "In many respects, the African hunger problem is imported. In colonial times traditional African crops such as millet, cassava and sorghum were abandoned in favor of cash crops such as tobacco, cotton, coffee and tea, which exhausted the soil and resulted in destruction of the forests." She was recently in Alberta taking part in the Ten Days For World Development programs by the churches. And Edward Saounia further declared: "We thought Africa would be a simple field where modern technology could be directly effective. We made a grave mistake. The frailty and complexity of the African agricultural ecosystems, already unbalanced by the extension of the colonial plantation economy, could not withstand indiscriminate infusions of costly modern technologies." Saouma made this declaration at a ceremony at the University of Florence, Italy, where he received an honorary doctorate in tropical and sub-tropical agriculture.

And, added Ogana, "foreign experts insisted upon clean farming in straight rows. People adopted Western methods of making a field look as neat as possible. The result was destructive to much African soil. "Now, Africans are learning to put back leaves and grass so humus can build up." Her conclusion is African problems need African solutions. They will do best if the wealthy countries will simply send them the money and allow them to spend it themselves. She must have been listening to Quebec! She admitted many African governments had their priorities reversed. They have spent billions on prestige airlines and prestige hotels for tourists to create urban jobs. It was only when people began starving they saw the light and went back to agriculture.

The Africans have had to put up with the posturings of too many intelligent fools. One of them is a professor at Israel's Ben Gurion University at Beersheba. Reuven Yagil postulates that African agriculture can be saved by the introduction of the camel. He has made an intensive veterinary study of camels and says they will fill the gap. But the whole thrust of his claim is that the Israelis are better able to handle camels than their nemeses, the Arabs. Camels can ride out the drought by going 23 days without water and Yagil figures that's the answer.

Eugene Whelan, Canada's former minister of agriculture, had a scheme that many persons laughed at but seems more sensible than Yagil's. He wants to stop the march of the Sahara Desert by dropping tree seeds encased in plastic darts from airplanes. The dart would shatter on impact and deposit seed on the ground. When the first moisture breaks the seed cases they germinate. This technique has worked on other similar terrain and the techniques for planting crops in semi-arid lands could be used in Western Canada. Whelan also says the legume tree should be introduced into the desert areas of Africa. It puts nitrogen fertilizer into the ground and this provides nitrogen for pasture grown between the tree rows. The legume tree grows at a phenomenal rate: eight inches in diameter in six years.

Whatever kind of agriculture is being carried out in the developing world, a higher percentage of people are being better fed for the first time in 40 years. Edward Saouma said the fifth world food survey has revealed there is a basic nutrition gain, i.e., a lower percentage is undernourished. This is encouraging news considering the rate at which world population has gained.

SOUTH AFRICA – *The Calgary Herald* – August 23, 1973

History will be made in the near future when the first shipment of six Hereford bulls leaves Canada for South Africa. They were selected by a group of breeders who were here at the end of June under the auspices of the Canada department of industry, trade and commerce.

While in Johannesburg earlier in the month I contacted one of the mission, Arnold Haden, administrator of Willow Grove Herefords, Balfour, Transvaal, for his impressions of Canada. "We were up to our necks in Herefords," Mr. Haden said. "We saw some excellent quality cattle. The breeders in Canada have the message that the big growthy animal is required and they are producing just such an animal." Mr. Haden was hugely impressed with the females at the S.E. Harwood farm near Calgary. They have a plentiful milk supply. He said polled Herefords were away ahead of anything he had seen elsewhere.

Although we regard South Africa as a "hot country," it really isn't that hot. They had freak snowstorms in the extreme south while we were there and had to send out the army to make several helicopter rescues in the mountains. On the other hand, Mr. Haden found Regina and Toronto summer heat of 100 degrees unbearable. Eating supper at 11 p.m. in daylight in Alberta was somewhat disconcerting.

Tim Henderson, past president of the South African Hereford Association, Mooi Farm, Natal, said an increased demand for beef in his country was the reason they looked at Canadian breeding stock. Evidently there are more and more cattle being finished on grain in South Africa. John Van Breda, the association secretary, who hopes to come to Canada on the next trade mission, says South Africa lays claim to the largest herd of registered Herefords in the world. This is the Jarrosson Estates of Freiburg in north-western Cape Province, the Texas of South Africa. On 400,000 acres there is 1,000 head of registered Hereford females and a 70,000-head herd of commercial cattle of mixed breeds.

Of all the sights that impressed the African delegates, the summer pastures were the most impressive. This, they figured, puts Canada on top as far as the cattle business is concerned. They should come around next winter—or have even visited Grande Prairie in that snowstorm last week.

Incidentally, they have a little different way of registering cattle in South Africa. They register all that are eligible and then send around a society official to inspect each calf before final acceptance. This must be quite a chore. One of the last things I expected to see in Johannesburg popped up while driving along a busy street with John Van Breda. It was a Lincoln Continental with a British Columbia license plate.

SOUTH AFRICA - *The Vulcan Advocate* - October 28, 1987

Once again an argument has developed about whether prairie-grown bread wheat has been, or ought to be, sold for consumption by blacks in South Africa. There doesn't seem to be any argument about whether Canadian wheat should be sold to the blacks of Nigeria, the whites of the Soviet Union, the yellows of China or the browns of India. It's only the South African blacks who should be deprived. A year ago External Affairs Minister Joe Clark ordered the Canadian Wheat Board not to make any more sales as one of the sanctions against South Africa to force the government there to revoke more of its apartheid laws. At the time Clark gave the order, a sale had just been completed to South Africa, which at that time had been suffering from drought and still has not recovered from its effects.

The wheat was imported to help the blacks spin out their supplies of white corn from which they make their traditional "mealies." It must be white corn as they won't eat yellow corn. White corn is not as available as yellow corn on the world market. But Clark was quoted as saying in 1986: "There will be no more sales." When he was challenged to say whether Western grain growers would be compensated for any "lost sales" Clark brushed that off with: "You are talking about something that hasn't happened yet—and isn't likely to happen." Only a year later, despite Clark's pompous statement, *The Saskatoon Star-Phoenix* reported the sale of a million bushels to South Africa. All export sales of wheat are priced by the one-desk Crown selling agency, the Canadian Wheat Board in Winnipeg.

There are several different ways of effecting the actual sale and movement of the Canadian wheat to the foreign buyers. The board's five commissioners can go abroad and beat the bushes for orders. Having received an order, the board can turn it over to one of its 25 agents, which have established credit ratings with it. The agent will receive a fee for doing the work of getting it offshore.

If the customer wants credit, the CWB can't give credit as it has no funds. Its mandate is to sell for cash to any buyer who can meet its price. If credit is required to make the sale, the CWB will call in the Canadian government—and the government will decide if it is politically wise and feasible to back a credit note. This is called a govern-ment-to-government sale and a large percentage of wheat is sold government-to-government nowadays.

The other way a sale can be made is through one of its accredited agents such as Cargill, Bunge, Continental Grain, Andre, or a number of private Canadian companies. A company with a tender in its possession goes to the CWB and asks for prices. If the CWB gives it a price on which it can make a profit, a deal is made and the grain goes out in the normal way. On the rationale of customer confidentiality the CWB will not reveal anything about its dealings, its agents, or the prices for which it sells the grain.

Other reporters got on the Saskatoon paper's story for a follow-up. One of them called the CWB and asked: "Why did you sell wheat to South Africa in contravention of Hon. Joe Clark's orders?"

The CWB replied: "First of all, we never received a written notification from the government that it supported Joe Clark's decision. Secondly, we won't confirm nor deny a sale has been made to South Africa." The CWB was right in adopting this stand because it did not sell the wheat to South Africa. Dick Dawson of Cargill Grain Ltd. said the grain was sold to a big international corporation, Andre Grain Co., based in Switzerland. But Andre Matraux, a representative of that company in Winnipeg, said: "We didn't sell to South Africa."

The CWB, Dawson and Metraux were all telling the truth to that dumb reporter. The scenario is that CWB likely priced the wheat to the Andre company and the Andre company could have sold it to another company or nation, which, in turn, could have forwarded it to South Africa once it was on the high seas. This once again proves what everybody but Joe Clark knows: That sanctions don't work. If South Africa wants wheat it will get it if it has the money. And 95% of Prairie grain growers, who are scratching for a market of any kind, will agree.

SOUTH AMERICA

Every tourist's aim is to attend upon Machu Picchu—and farmers are no exception. A group of Canadian farmers towed Schmidt along with them. This story enticed a whole new crop of readers and they stayed with his column. During his travels, Schmidt discovered that Canadians have cousins in South America resulting from the dispersion of European emigrants into the new world. Curiously and without much intervention over hundreds of years, the South American cousins had adopted similar farming method to Canadians.

He also uncovered the secret of why Argentines and Brazilians eat two hundred pounds of grass-fed beef per capita annually and why Uruguayans eat five hundred pounds. Canadians, on the other hand, consume only 60 pounds of grain-fed beef. On a visit to Colombia in 1977, he went looking for farmers picking coca leaves for the drug trade, but the only drugs he could find were coffee beans. In deference to his more sensitive readers, he provides a disclaimer for the revelations about the darker side of South America countries and suggests they may wish to simply skip over them.

MACHU PICCHU, Peru - *The Calgary Herald* - February 20, 1971

It was very funny to see our group of Canadian farmers roll out of bed at 5:30 a.m. at Cuzco, complaining about being awakened by a rooster which had begun crowing at 4 a.m. So few Alberta farmers own a flock of chickens now that a rooster crowing is a novelty—especially in a city. The purpose of the early morning awakening was to make a 70-mile train trip to tramp through the ruins of the ancient Inca city of Machu Picchu, Peru's chief tourist attraction. A certain amount of mystery surrounds the location yet.

Placing the heavy stones of the temples, homes, farm terraces, towers, fountains, jail, trapezoidal and rectangular niches and doorways in this city built 2,000 years ago would test the skills of modern engineers and builders. The Incas had no known means of leverage, not even the wheel. How did they get the stones there? How did they manage to cut them so accurately that the walls stand up without the use of mortar? Did this work require generations to complete? Was it done by slaves? What happened to the architects and the plans? Archaeologists now believe the city was inhabited 70% by women. It could have been a haven for the Incas' sacred vestal virgins. One might visualize it as the last stand of the first women's liberation movement in the world.

Getting to the ruins is half the fun. The narrow gauge railway rises to an altitude of 12,500 feet out of Cuzco. It then drops down to 9,000 feet through a series of valleys, whose rivers make up the headwaters of the Amazon. The railway continues out to the jungle city of Santa Ana. To climb 2,000 feet out of Cuzco requires four switch-backs. Hillside slums have grown up alongside the switchback tracks. It seemed the track was running through the people's bedrooms. Hillsides, which no one else seemed able to make use of, are the favourite haven of squatters. All cities have hillsides built up of squatting families, who appear overnight and erect a few rudimentary poles over which woven bamboo mats are hung.

At stops along the way, peasants appeared along the platform with unappetizing-looking fruit—peaches, apples, and mangoes—for sale. But one real delicacy was hot corn on the cob; only the kernels were two to three times the size of the Canadian sweet corn and much juicier. Corn and potatoes of infinite variety and type grow in these mountainous regions and have drawn botanists from all over the world to classify them.

Sanitation among the natives, who operate roadside stands and in some of the outdoor markets, is very rudimentary. While government agencies are working hard to improve health standards, the visitor is repeatedly warned about the "bugs" to be picked up from native-grown produce. Water is the main problem.

After the three-hour run to Machu Picchu, there is a 20-minute ride in a jitney up a precipitous 2,000-foot slope with 13 hairpin turns in the one-lane road. The Peruvians are good engineers and have built the road so a jitney can make the ascent without changing gears. This is a good thing because the jitneys creep up in bull-low gear and there is no other gear to shift down to should they power out. For me, the trip back down to the Machu Picchu station was a white-knuckle ride. It was a hurried trip of about 10 minutes over a road that was then made slick by a heavy rain. But the Indian driver was more sure of his, and his vehicle's, capabilities than I was. After all, who was I to ques-tion the capabilities of a person whose ancestors have traveled and lived a cliffhanging existence for 20 centuries?

The whole exercise of making the trip to Macchu Picchu made one pause to reflect the course of civilization. Here were people living in what our civilization considers a primitive fashion. Families were tending a few nondescript looking sheep, cattle and llamas. Not a few of them were carrying transistor radios, their main link with outside civilization. CBC-Spanish short-wave programs are heard here. And a lot of cheap North American "culture" is invading the primitive areas. The big question is: Which way is their civilization proceeding—up or down, backward or forward?

SANTIAGO, Chile - *The Calgary Herald* - March 2, 1971

The other day in Lima, Peru, a Canadian and an American businessman were discussing their various duties. Both had been working in and around South American countries for a number of years. The Canadian, from Prince Edward Island, had been here eight years. He laughs uproariously at Canadian ignorance of South America. "I have become so used to the easy pace of Latin living I'd hate to have to adjust to the North American rat race again," said the Canadian.

"I'm in about the same boat," said his American friend. "The more leisurely pace suits me perfectly. The Chileans have their methods of getting things done here—but they get done." Sometimes the pace is so slow (and frustrating) that to their more active North American counterparts it appears to grind to a complete and irreversible halt. The only way to cope with this Latin way of doing business is to shrug and accept it. Acceptance may even mean never accomplishing a project, frustrating as that is. Here are a couple of examples of what I mean.

In Cuzco, Peru, I met a government veterinarian who was anxious I see his brother, a Ph.D. from Harvard University, who was a top planning officer in the Peruvian government planning department in Lima. He had done some work on oil agreements with the United States and his brother suggested I might learn something of agrarian reform from him. A phone call to him revealed our schedules conflicted, but rather than abandon the interview he suggested I come to his office at 8 o'clock on a Friday night, after he had concluded some meetings. He gave me the address of a government building four blocks from the hotel and said anyone would direct me to his office when I arrived.

I arrived at the appointed time and place to be confronted by a non-English speaking guard at the door who stoutly maintained the boss had left and wouldn't be back. There were lights upstairs but in no way would he let anyone investigate if the boss had left a message or if he was still there. That was the last I ever heard from this gentleman. When I mentioned it to my American friend, he laughed: "They'll make all kinds of extravagant promises, these people, but break them—even with their best friends. You'll see it happening again and again."

And I did—a couple of days later. At this point I should mention Chilean Immigration officials, both going in and coming out of the country, insisted upon knowing the exact nature of my job and the name of the paper for which I worked. While drinking our way through the Conchay Toro Winery on the outskirts of Santiago I struck up a conversation with an English-speaking Chilean who was there with his family. He turned out to be the boss of the government's commercial textile importing corporation.

When he learned I was an agricultural journalist, he suggested I might be interested in the agricultural reform policy of President Salvador Allende. "Chile has a story to tell the world," he said. He wanted Canada to know the story. He journeys to Japan several times a year via CP Air through Vancouver. He volunteered to have someone from the agriculture department get in touch with me at the hotel the next afternoon to tell me the story. I agreed to have a translator on hand. Next day at noon a message at the hotel said: "Avisan sue a las 13:00 horas vendra una persona del Ministerio de Agricultura a hablar con usted al hotel."

I waited in vain for the man from the agricultural department to show up. At the end of three hours it was time to board an Aerolinas Argentinas jet for an airlift over the massive Andes to Mendoza, Argentina. The only recourse the people of Canada now have to the truth about Chile is through "Operation Truth," which has been started by President Allende. He thinks it is about time to end the lies spread about Chile by international agencies like the CIA. Evidently not too happy with Time's recent story, he has organized a corps of Chilean journalists to tour the world to tell the true story.

My friend in the winery was evidently in accord with his president's dictum that the truth be spread to the world, and was doing his part to help. The thing is: how does a country sell itself, when it has a socialist government run on democratic lines with a Marxist president? The land reform was begun by the previous Christian Democratic regime of President Frei. Graft entered the picture when government supporters began unloading their holdings and taking the money out of the country.

Although Allende is a professed Marxist, his support has come from a Maoist element and land reform is being carried out now by the agriculture minister who had a four-year apprenticeship in Cuba. There is more to it than merely blatantly seizing land and dividing it among smallholders. It is a redistribution of wealth. Many of the moneyed people have fled the country. Many of the wealthy—who are in the minority, of course—have come to the conclusion the end is near as in On The Beach, and have gone on a wild spending spree which has caused, by official admission, a 40% inflation. One person of my acquaintance who returned to the country after a 10-month absence said food, drink and other living costs had risen 100%.

Chile has problems, but the official version of them will have to be told in Canada by those Chilean journalists sent by Allende. The reason nobody from the office of the minister of agriculture, Hon. Jacques Chonchol, did not show up was because the minister was in real trouble. An opposition party charged he had broken the constitution by illegal takeovers of land in south Chile. Chonchol will be the second minister from the Allende government charged with violating the constitution. The minister of justice was charged with giving amnesty to some of his friends in jail.

It will be interesting to learn if Chonchol is impeached and shot.

CHILE - *The Calgary Herald* - August 23, 1986

In the world today there are 90 million unemployed and 300 million under-employed. The International Labour Organization estimates the world must create 47 million new jobs a year for the next 40 years. This is a mammoth challenge as the need for most of them will be in Africa, Asia and Central America. How is this formidable task to be accomplished?

Ivan Head, president of the International Development Research Centre, presented his answer to the 1986 Conference of Food Science and Technology in Calgary. In his preamble, the former Calgary lawyer and confidant of Pierre Trudeau, noted that the economies which look to be strongest in the world today have followed the agri-industry catalyst to modern industrial growth. "Of the developing countries which are exhibiting the most economic vigour, that pattern is still evident," he said, echoing what former agriculture minister Eugene Whelan said for a decade.

Sadly, many newly-independent countries have endeavoured to leap-frog this route to modernity; and have reaped disaster. "Fortunately, however, I discovered the lesson seems to be sinking in," Head said. "Inherent in this new policy thrust, I sincerely hope, is the realization that technology—to be effective—must be relevant. By relevant, I don't mean small, or low-tech. I mean that the technology must be related to real demand, not the reverse."

Unfortunately, not all the developing countries want to go the agri-based route. There must be a reason for this—and Head, who has his finger on the pulse of the Third World in his IDRC position, discovered it. Those countries were swayed by arguments of Raul Prebisch, the brilliant Chilean economist, who died in May. Prebisch had argued that the prices of primary commodities, relative to those of manufactured goods, would always decline. Prebisch contended that the demand in the industrialized countries for tropical agricultural products was inelastic while the market in the developing countries for manufactured goods was highly elastic. Accepting those premises, many developing countries, particularly in Latin America, adopted a capital-intensive, import-substitution strategy. The one country where it has been successful, and strikingly so, is Brazil.

An unhappy witness to truth of the Prebisch doctrine was Tanzania. An outraged Julius Nyerere again and again would compare the immense gap that increasingly widen international markets between the price of manufactured goods—tractors were the example that Nyerere usually cited—and the price of the agricultural commodities such as sisal or cotton that the developing countries sold in order to earn foreign exchange required to pay for their needed imports. That widening margin was one factor that contributed to the debt burden.

But Prebisch was only half-right and it's going to take much persuasion to convince many of the countries they are wrong-headed in bypassing the agriculture phase of development. Of course, it is hard for any government to resist the pressure of large numbers of unemployed for jobs—jobs of any kind—than to get the country switched onto an agricultural economy which takes time. But, concluded Head, "it is clear the most successful strategy is one which focuses on food processing and distribution"— a policy which will redistribute money into the pockets of the poor to buy adequate food.

SANTOS, Brazil - *The Calgary Herald* - March 17, 1971

The columns written from South America have pretty well parroted a mass of information a group of Canadian farmers and ranchers was given during a month-long birdseye view of Peru, Chile, Argentina, Uruguay and Brazil. I must point out before concluding the series that most of these were fleeting impressions. We traveled hard and didn't stay anywhere long enough to become experts. Parroting has its hazards, I learned after interviewing a blue parrot in a zoological garden here. The parrot told me not to write everything I was told about Brazil.

The thing I find hardest to believe about Brazil is the death squad, a vigilante gang among police, which has undertaken to clean up crime by summarily executing criminals. Although police have up to now denied its existence as mere newspaper speculation, 15 policemen have been arrested and charged with murder.

What is appealing enough about this system of justice that it could well be imported in Canada or the United States is that the 1,000 who have been executed since 1958 comprise mostly habitual criminals, murderers who have been freed after 15 years and known drug traffickers. There is no official death penalty in Brazil and the police who have been taking the law into their own hands know they won't be hung if they are caught. They have found more than a little public support for this type of justice. Because police investigation of other police is usually perfunctory and because no murderers have been hung in Canada recently, there is no real reason such methods couldn't be tried here, within limits, it was suggested to us by Brazilians.

My notebook shows the South Americans solved many other problems with answers that have eluded us. Many of the big cities have been built inland several miles from the hot, muggy coast. Sao Paulo is one of these—some 2,500 feet up and 50 miles from the seaport city of Santos. The climate is much cooler. Santos is a city of 300,000 and is reached by a fantastically engineered freeway that winds along escarpments, virgin forest and through tunnels. Canadian engineers hack and blast wide enough rights-of-way to accommodate four lanes. Brazilian engineers adopt an easier course of building the two right-hand lanes at different elevations and roadbed than the left-hand lanes. Often it's impossible to see the other lane of traffic. However, the grades are better and more cheaply constructed.

Brazil claims to have no racial problem although 40% of the population is illiterate and many of the illiterate colored people are flocking to the big cities from the northeast. It is the frankly-held view of the white people in the south that Indians and Negroes have caused no problem because "they know their place" and haven't created the same difficulties as in the United States and South Africa. In the north there is a wide variety and mixture of races, colours and creeds. Uruguay frankly admits it shot off a tribe of 16,000 Indians it didn't think could be "civilized." The feat is recognized by a monument in Montevideo. Aggressive ethnic groups are given wide scope. For instance, Sao Paulo has the largest Japanese population of any city in the world outside Japan. Immigration of Japanese families for agricultural pursuits is encouraged. The Japanese control 70% of the country's agriculture and most of the fishing fleet. Cattle ranching is the only enterprise not controlled by the Japanese.

Parking problems, in some of the big cities, are solved simply by parking cars on the sidewalks. It's not unusual for the pedestrians to be sharing the walks in front of apartment buildings and downtown businesses with half a dozen cars. One of the treats in semi-tropical or tropical countries is eating dead-ripe pineapples, which are big and cheap. No matter how one tries to enjoy this culinary delight in a civilized way, the juice gets all over. The only satisfactory way to handle the matter is to take it into the bathtub. And, incidentally, those stories about wine being cheaper than water are true. In cities where domestic water supplies are unreliable and bottled water had to be bought we made this important discovery.

Fortunately, people don't take politicians as seriously as we do in Canada. "We wake up some morning and find another president has been sent packing. Somebody new is in office—and we don't know how he got there," was the way a woman tour guide in Buenos Aires described the situation. "Would you kiss Prime Minister Trudeau if he showed up here?" she was asked. (This was before the PM's marriage.) "What? On his bald head?" she retorted."

There is a great lack of understanding between the countries of the two great continents. Lack of communication, language, trade, distance and opportunity for exchange are the main reasons. An instance of this occurred in Campinas, Brazil during a luncheon at a gaucho restaurant. About 20 from the city council and engineering staff were having a luncheon at an adjoining table specially set up for them. The boss man came over and extended greetings in English to us. An accordionist and a saxophonist came in to serenade both groups. The restaurant owner said the duo would play in honor of the Canadians. They played Anchors Aweigh, followed by a barely recognizable version of La Marseillaise. Trouble was nobody from our group could think of an internationally recognized Canadian tune for them to play.

BRAZIL – *Fort Macleod Gazette* – February 5, 1992

Ralph Jesperson of Stony Plain, Alberta, the ebullient president of the Canadian Federation of Agriculture, was telling us not long ago at the annual meeting of Unifarm in Edmonton about that big environment meeting in June in Brazil. It's probably going to be the biggest the world has ever seen. Up to 30,000 from around the world have signified their intention to attend this meeting to sit out in the middle of the rain forest to see if it gets sawed down around them. They will discuss preserving the environment of the planet and its natural resources (including the rain forest and rare species of plant and animal life found therein).

This will be the trendy place to hear the latest ecological buzzwords such as biodiversity, high tech, sustainability, economic earthquakes, global warming and greenhouse gases. I'm not exactly sure who is organizing this junket or if it is a matter of spontaneous combustion. I don't know how long the 30,000 are going to sit knees-to-knees in think tanks. Brazil will be told its policies are all wrong; that it shouldn't be cutting down some of its rain forests to feed its people. But rather than just a mere 30,000, I would like to see a million people go to Brazil and travel the length and breadth of that vibrant country for three or four months. They would have a great deal to learn. I have only been there twice for short visits but those visits have whetted my appetite to learn more about its vitality.

It is necessary to go and look over the entire country first before making any judgments exported here by dissidents. Before becoming a critic, let each of us rectify our own national mistakes. In the 1970s, the national policy of Brazil was to raise the population to 900 million from a population then equal to that of the United States or the Soviet Union. That was a political objective. No measure of common sense could dissuade Brazilian politicians from going ahead with this foolhardy policy, the result being the necessity to clear more land to grow enough food for the multitude. No Canadian can criticize until that Canadian persuades the Quebec government to abandon the increased baby bonus, also urges the Canadian government not to fill up this country with Third World immigrants. Other political experiments such as medicare, supply management, unemployment insurance and bilingualism threaten to bankrupt this country any day now.

Let a million people go to Brazil and they will find a populace whose colors range from ash blonde to blue-black getting along peaceably with no constitutional wrangling and posturing. Agriculture in Brazil is a dynamic industry and is the envy of Canadians on these prairies which are dried out most of the time. It is a country full of poisonous snakes and the ability to produce two or three crops a year. It is filled with wondrous and imaginative engineering projects. The crush of humanity on the streets of Sao Paulo will scare the claustrophobic and that same humanity is generally poor and will rob you blind; many will spend a year's wages rigging up a float for the Carnival in February.

One of the learning experiences in Brazil would have to be jungle politics as practiced by the Kayapo Indians and the 90,000 rubber tappers. They have even convinced many people throughout Canada this mild winter is the result of rainforest logging. Pretty astute these politicians. The rubber tappers have demonstrated politically the difference between ecology and the economy. The tappers live in paradise in the Amazon rain forest and belong to a union whose leader, Chico Mendez, started a campaign to persuade the government to take action to save the wild rubber trees from which they gather latex (for the making of natural rubber) from destruction by farmers and ranchers.

In 1988 Mendez was assassinated. To atone for the crime, the Brazilian government set aside a 2.5 million acre rubber tree reserve (that's about twice the size of Prince Edward Island). But alas, the tappers saved the trees but there is now no market for latex. It has been replaced by superior synthetic materials. The tappers can make only about $30 a week. Therefore, they, themselves, are clearing patches of rain forest (alongside the ranchers and farmers) to grow food to maintain their families. The alternative is to go and starve in the city favellas.

Things change. The green movement of North America and Europe sees them as militant ecological guardians of the rain forest but, in fact, they are also clearing it for economical purposes.

COLOMBIA – *The Calgary Herald* – January 28, 1977

The original purpose of the trip into the Sierra Nevada Mountains of Colombia during the Christmas holidays was to visit a coffee plantation. Getting chewed up by mosquitoes and blackflies was an unexpected and unwelcome side-effect of the trip. Although I kept my eye out for monkeys and boa constrictors, the most surprising sight I came across in the middle of the jungle was a poultry farm with 3,500 laying hens.

The trip was made with a local Land Rover jockey and travel man named William, who said the plantation was near a small village named Minka, about 90 kilometers from Santa Marta, a Caribbean resort visited by Canadian tourists on an organized basis for the first time this winter. Before taking off, William topped off the gas tank with 27-cent-a-gallon (repeat for the benefit at gasoline-tax-hungry politicians, 27 cents) gas. We were then ready for a jolting trip over a road more rugged than any B.C. logging road but he had to lock in the four-wheel-drive for only about six kilometers. Although this was a main road of commerce used by buses, trucks and taxis, its original purpose 25 years ago was as a haul road in the construction of a television tower high in the mountains. There the road stopped. It was not practical to take it any further into the mountains which rise to 19,000 feet.

This is a country of contrasts. About 60% of the people cannot afford TV, the original purpose for which the road was built. This is a country where it is cheaper to carry people by air than build roads through the fearsome mountain terrain. By Canadian standards, public transport is dirt cheap. City bus fares are nine cents but there are lower class buses where the fare is four cents. Taxis are comparably cheap. In Minka we saw women doing their washing in a stream and men toting wood home on their backs. However, there was a modern taxi in the village which had brought in some fares from Santa Marta. One day in front of a middle-class Santa Marta home I saw a taxi pull up and deposit a mother and her son who had been shopping. The son was carrying a live chicken for the evening meal. Another attraction in Minka was the village dude galloping about with a scarlet saddle blanket and silver spurs.

An attack by swarms of blackflies and mosquitoes (such as we would experience in Canada in July) left me with swollen arms and legs for a couple of days. The bugs really enjoyed a change in diet. I am told oil of marijuana is a good mosquito repellent and I must try it some time.

Foreign visitors to Canada are usually taken to the showplace farms. William was no different. He took us to 2,500-acre Victoria Plantation owned by a German named Weber, who had been there 30 years and was considered a rich man. With a 7,000-bag coffee crop this year he should make money. In a sense, this was not a typical plantation as the average Colombian grower has 10 acres. The 2,500 acres was spread over the wildest, hilliest, jungle terrain imaginable. The coffee bushes were planted over the thickly wooded mountainsides for a combination of shade and sunlight, which gives Colombian coffee the best flavour in the world.

The beans were all picked by hand and how this is accomplished without the aid of parachutes or skyhooks on the steep hair-raising precipices I don't know. The beans are red and resemble cranberries. The pickers fill 70-kilo sacks and load two each on donkeys which carry them back to the main headquarters where the husks are removed under water pressure, and dried before the beans are shipped off to market. We came upon a string of loaded donkeys on the side road to the plantation. One had slipped its load and the bags were hanging underneath it. As the driver was nowhere in sight, William and several of the men proceeded to unlash the sacks and load them onto the Land Rover. The sacks were almost as heavy as the donkey carrying them.

The coffee bushes, which require three years to reach maturity, are started in greenhouses and transplanted all over the place on a helter-skelter basis not in rows in fields. There appears to be plenty of room for expansion as two-thirds of the mountain and jungle of Colombia is not populated.

We continued on up over the cliff-hanging roads to a lunch stop near a stream overlooking a wide valley and a panoramic view extending back to the Caribbean. High on a mountain in a jungle clearing was a poultry farm. There were about 3,500 hens and a proper complement of roosters. It was a complete surprise and an unusual place to produce eggs. William said the reason the chicken pens were built that far back in the jungle was cheap land. It is cheaper to haul feed to the hens there and haul the eggs to Santa Marta over the terrible road than buy the extremely high-priced land closer to town. The land in that South American country would be regarded as too high-priced to farm in Canada. But it is farmed and somehow the farmers make a living. This may be the answer to why foreigners are willing to pay what we consider high farmland prices in Canada—much higher than we think profitable.

Incidentally, the Colombians eat a lot of iguana eggs. William described them as larger and meatier than hen's eggs. On the return through Minka, William treated all who could handle it to a drink called aguardiente, tipped back neat. It is a sugar cane-based liquor with a smooth liquorice flavour.

NICARAGUA – *The Calgary Herald* – October 6, 1986

Alberta Farmers For Peace, a non-governmental organization headed by Irving Bablitz of Bruce, joined with the Council of Social Affairs of the Roman Catholic Church to bring Justinian Liebel of Managua to Calgary. The purpose was to encourage Albertans to continue to support the work of both organizations among the new farmers in Nicaragua. In the past, Albertans have been generous with cash and farm implements and groups of farmers took time off to go to that Central American country to assist the Sandinista agrarian reform.

Liebel is a former priest from the U.S. midwest who has spent 20 years as director of CEPA working among the Miskito Indians of Nicaragua. CEPA is the Center for Education for Agrarian Reform, an agency of the Roman Catholic Church. The Alberta Farmers for Peace has supported it on several projects. Brick-making machines and Alberta goats were received by CEPA.

As a farm writer, I was interested in the fact that *La Prensa*, a newspaper which is an institution in Nicaragua, had been shut down by President Daniel Ortega in late June. This newspaper had supported Ortega's Sandinistas in overthrowing the Somoza dictatorship and now, ironically, has become its first victim. I asked Liebel, a Sandinista sup-porter, to retail *La Prensa's* sins. He obliged with a long, involved, convoluted political explanation, which revolved around the fact *La Prensa* reported everything, even the sins of the Sandinistas and the activities of the Contras, who are carrying out a counter-revolution with $100 million in U.S. government support. "We look at the American support to the Contras as a Reagan declaration of war," he said, and added: "When the U.S. stops this war we'll let *La Prensa* publish again." No more will the good works of Alberta Farmers For Peace be chronicled in *La Prensa* until then.

The ironic part of his explanation was that for every reason he gave for closing *La Prensa*, the same reasons could apply equally for closing *The Calgary Herald*— should the government of Canada take the same measures. But Liebel wouldn't buy that. "I would never advocate *The Herald* be closed," he said hastily.

No. As long as there is someone at *The Herald* to write about the evilness of the Contras and the evilness of the Americans who support them, he wants to see it around. At the conclusion of the interview, Liebel offered to pray that I would get the story across to *Herald* readers.

I was more interested in praying that *La Prensa* be reopened for the sake of Nicaragua's future and that its farm writers be put back to work. The people of Nicaragua have reaped immense benefits from the diligent work of its editorial staff. They have had hundreds of grievances redressed with *La Prensa* help.

When the people weep, the press weeps with them. But when *La Prensa* wept at grievances to the people caused by the Ortega government, it wept alone. The press is the conscience of those who handle the destiny of the common man. It did much for Liebel and his people intellectually. But when the chips were down his intellectuality disappeared. Of course, there isn't too much intellectuality coming out the barrels of guns. I pray that *La Prensa* will be reopened because individuals cannot take the place of institutions. I pray that, before it is too late, Alberta Farmers For Peace will take to Nicaragua with them a copy of a homily written by J. P. Curson: "Because it is the advocate of society, therefore of the people and of domestic liberty, I conjure you to guard the liberty of the press. It is the great sentinel of the state, a great detector of public imposture. Guard it; because when it sinks there sinks with it in one common grave the liberty of the subject and the security of the state."

Great God, it's a topsy-turvy world you've given us when an ink-stained wretch feels compelled to preach to a priest and a priest feels compelled to wage war!

BUENOS AIRES, Argentina
– *The Calgary Herald* – March 8, 1971

The world demand for meat now exceeds the supply. In view of this state of affairs, one would expect Argentine farmers and ranchers would be rubbing their hands with glee as their country is one of the world's largest meat exporters. Meat earns this country more than $500 million annually in foreign exchange. But the producers are not happy. They have become enmeshed in a trap from which the government is clumsily trying to extricate them. The trap is an outmoded packing industry, which cannot or refuses to modernize. The ranchers have kept pace but they are stymied by a shortage of government and bank credit. Also two years ago they were hit by a land tax which was imposed on top of all other taxes.

This latter story has a familiar ring in Alberta and British Columbia. It would almost seem Premiers Harry Strom and W.A.C. Bennett had been trading notes with El Presidente Marcelo Levingston of Argentina.

For 75 years into the 1950s Argentine meat packers have been able to capitalize on the frozen meat trade in Europe. They were geared to freeze the meat and ship it in refrigerator ships. But in the 1950s a change in buyer preference occurred. The buyers don't want the whole frozen carcass any more; they want chilled and frozen—roasts and steaks—and canned and cooked meats.

The huge meat export industry is dominated by three companies, which say they don't want to retool unless they can buy cattle cheaper from the farmers and ranchers. This means the producers must turn gradually to grain-fed beef rather than traditional grass-fed beef, which take more land and time to produce. Last year the big three were taken to court by nationalistic Argentina trust-busters on market manipulation charges. Much nationalistic feeling was stirred up and they were convicted. An appeals court threw out the conviction and said it could see no market monopoly. In view of the fact the companies were losing money, they have retaliated with counter-measures which threaten 30,000 jobs. Last October they all suspended sales to export markets. Closure has created great socio-economic problems among people in the area. The closures were described to me as "shocking."

The crises in the export business will not be resolved, the big packers say, until the price they pay for cattle to producers goes back down to about 12 cents a pound. Domestic consumption has pushed it up to 16 cents. Some of the smaller packers with modern machinery say they can make a profit on the export market buying at 16 cents. However, the government is looking at that $500 million earned in foreign exchange. The ranchers fear the government may force them to divert cattle into the export trade at a lower price. Consumers fear if this happens they will have to pay more for their meat. They are kicking like steers now at having to pay $1 in restaurants for a sirloin steak that cost them only 60 cents a year ago.

MONTEVIDEO, Uruguay – *The Calgary Herald* – March 11, 1971

The Montevideo stockyards are run on a vastly different method of operation than the Calgary yards—or the auction market at Red Deer for that matter. There doesn't seem to be any organized system of corrals or sorting pens. The operation is carried out in a large open field by cowboys (or gauchos) on horseback. From there the cattle are hazed onto the weigh scales.

Our group of Canadian farmers and ranchers, who visited the operation on the outskirts of the city, found six firms located half a mile or so from each other. The cattle had been trucked in, trailed in or brought by rail overnight. Sale was by private treaty early in the morning, presumably soon after the cattle arrived at the yards of the commission firm of the farmer's choice, or the sale could have been carried out before they left the farm. At the Calgary yards, the cattle are sold to the highest bidder by auction.

When we arrived at the yards at 7:15 a.m., on a bright but slightly cool day the gauchos were holding cattle in small bunches. The gauchos don't use whips. They use two-inch-wide leather straps on the end of a riding crop that make a lot of noise but don't damage the meat. The cattle were identified with the number of the packer who bought them and then trailed several miles to the respective plants. There they are held for three days (although not starved) to make certain they are under no stress before slaughter. Cattle killed under stress yield tough stringy meat.

There are seven packing plants in the city and they supply both the one million population and the export market. Other cities in Uruguay slaughter only for the domestic trade. For the domestic trade, meat is held 48 hours at room temperature and sold without benefit of refrigeration. Much is sold on outdoor street vendors' stands, cut to order for the housewife, minus or plus a few flies. The meat for the foreign trade is chilled, frozen or canned. The nature of the processing by the Montevideo plants makes for the widest variation in quality of cattle going through the yards. Everything was on hand from a 10-year-old steer that had been used for pulling a plough, to a bunch of old cows with cancer eye, to some prime Charolais and Hereford steers.

Of note here is the fact Uruguay has been importing Charolais breeding stock for some years (they already have foot-and-mouth disease so there was no difficulty in importing from France). Another note: When I mentioned the fact that Canada ate all its domestic production of beef and imported additional supplies, a Uruguayan said: "Then you don't produce enough beef cattle. You should produce enough to be able to export at least one-third of your production."

Back at the stockyards, there were a great many skinny, runty little cattle of both sexes for boiling down into canned meat. They weighed 400 to 600 pounds and in Canada would be the kind sent back to farms for finishing on grain. When I expressed surprise at these kinds being slaughtered, I was told no grain feeding was done and they weren't worth wasting any more grass on. It was better to cull out the low-grade cattle and take two cents a pound than use grass that would produce a prime steer at 12 cents.

There is a law that a female cannot be slaughtered under age eight unless she cannot have a calf or something else is wrong with her. Another law is that although a farmer can handle his own cattle through the yards, the unions demand a minimum of three riders to be used for every head consigned. The labor unions in this country are evidently extremely strong. Our English-speaking guide, Mrs. Alma Cooper, had a special dislike of them: "They went to the United States to learn how to organize. Now they have a tight grip on the government. They're worse than the Communists. You can get rid of the Communists at the elections—but we're stuck with the unions," she said.

It is well nigh impossible for an employer to fire a union man after five years as the cost is too prohibitive. It is cheaper to keep him doing nothing than pay the high severance. I don't imagine the gauchos back on the farm are unionized. There don't seem to be many young riders. Reason is all the young bucks are heading to town to join unions and work in the relative security of the cities.

To maintain rural communities and prevent rural slums, the Uruguayan government has a scheme of tax reductions for the rural areas and tax incentives for persons starting industries there. One-third of the country's population lives in Montevideo. The scheme probably resembles Ottawa's rural development scheme—but bears little resem-blance to tax increases which have been levied on agricultural land on a provincial and municipal basis in Alberta.

The gauchos in Uruguay and southern Brasil (they don't like a "z" in their country's name here) are a colourful lot, dressed in wide-brimmed hats with a chinstrap, baggy pants, ornately decorated and big, sheathed knives at their backs. The knives are used for everything from digging stones out of their horses' hooves to slicing up their meat at the table. Their saddles have no horns, are cinched further back on the horse (almost in its middle rather than behind the front legs as on Canadian ranches) and the stirrups are further forward. Each gaucho has a sheepskin strapped over the saddle, ostensibly to give him a more comfortable seat. Our Canadian riders, like A.B. Murphy of Regina, found the sheepskin spread their legs too far apart for comfort. They could not get the horse between their legs and the saddle put them too far back on the horse.

In Brasil some gauchos were wearing mean-looking spurs. The rowels were four inches in diameter and built like a buzz saw. They were mounted on eight-inch shafts strapped to the boots, a la Calgary Stampede. I was not able to establish whether these were dress or ceremonial spurs or whether they used them on the range. To complete the picture, they knotted the horses' tails about the halfway point. This was partly traditional and partly to keep the tail clean. And they could ride like the wind, and sing and cook meat well.

URUGUAY - *Hanna Herald* - August 25, 1993

Most readers have heard about the programs on W-5 Eric Malling did on New Zealand and Saskatchewan on the CTV network. He compared their disastrous economic plight to our beloved Canada. Our problem is over 30% of our taxes go to pay interest and principal on the national debt. New Zealand had a total government debt of 62% of gross national product while Canada's combined federal-provincial debt is at 88%.

Uruguay (population 3.5 million) got itself into a similar situation in the 1970s and 1980s when the cost of its cradle-to-the-grave social programs spawned a burden of debt, which took that agriculture dependent South American country almost to the edge of bankruptcy. Last October, I started out on a holiday to Uruguay to find out how it saved itself and whether Canadians had anything to learn from their Spanish-speaking cousins' experience. However, as I outline in a previous column, this trip was aborted in midterm by lack of a proper visa. In the meantime, I have been able to learn what happened there, although not at first hand.

Uruguay is a 175-year-old democracy, a cut above most of the Southern Hemisphere banana republics run by tinpot dictators. Well before Mackenzie King took Canada down the road to socialism in the 1940s with the introduction of unemployment insurance, President Jose Batlle y Ordoronez had given his nation of middle class people a "model" welfare state. His innovations included the eight-hour day, minimum wage regulations, paid vacations, worker pensions, mothers' privileges, and free education. But like King, Judy La Marsh, Lester B. Pearson, Pierre Trudeau and some Red Tories, Batlle did not fund these and other programs on a realistic basis to create his paradise. This led to disaster and a military dictatorship in the 1970s.

Uruguay, like Canada in the 1930s and 1940s, was a society whose income was based chiefly on agriculture. In Batlle's time, the farmers' income was very large and paid the welfare shot. When this sector eventually ran out of money, so did the taxes, which funded the liberal social legislation. In the years after the Second World War Canada developed a more diversified industrial society and this masked the fact—a fact the politicians failed to realize—that the only resources to tap for the welfare state are farming, forestry, fishing and mining.

To be at all successful, a welfare state cannot be abused. But, as it has been in Canada, the Uruguay welfare state was severely abused. Women were retiring on state-paid pensions at 25 or 30 to look after their families. Men could retire at age 50 with full salary, provided they worked for 30 years. After death, their widows received 50% of their pensions. This meant a man could spend 20 years of his life not producing but drawing on the limited resource of taxes paid by farmers.

Uruguay thus went down the path to inflation and insolvency. The end result was the kind of government Cuba would provide. To keep themselves in power, the politicians bribed the taxpayers with their own money and government jobs. Even in the 1950s the bureaucracy had expanded to the extent a third of the population was on a government payroll or receiving government cheques. Government-run agencies like the post office were in the red. Free education proved to be no good as it produced more urban professionals than the nation could use and they provided a breeding ground for the subversive Tupamaros, who were the dupes of Castro and were the urban guerrillas. There was discontent everywhere to the extent the military had to step in and keep a lid on things.

The army's fight against the Tupamaros cast the generals in the role of death dealing politicians and resulted in an undeclared civil war—a terrible situation faced by the normally passive and non-violent Uruguayans. The army got rid of the terrorists but then the people had a terrible time getting rid of the army.

As a hard-nosed old curmudgeon, I'd like to report that the "good guys" got back into power again, threw out the "rascals" and everyone is living in prosperity today. But this has not happened yet after many years of trying. Or were they trying? My perverse nature leads me to believe there are too many urban voters who still want to see those government cheques—worth more than the taxes they pay—coming each month. What I really wanted to find out was how all the people survived when they were cut off government goodies and cheques and their taxes came down and they were able to put money back in their pants pockets to look after their own welfare.

I couldn't find an easy answer. After all, moving a nation from a Honda economy to a Cadillac economy and back to a Honda economy is not an easy proposition to handle. Let's send Don Mazankowski and Jim Dinning there to find out if it can really be done.

QUARTET OF TROPICAL COUNTRIES

A neighbour in Calgary told Schmidt about a Canadian financed foreign development project he had worked on in Belize. He suggested that instead of criticizing Canadian assistance to the Third World, the farm writer go to Belize to have a look at what Canada was trying to accomplish—and he did.

Wherever he looked, Schmidt found humans sculpting the world to accommodate their needs by reconfiguring various elements to rebalance nature. This, however, also results in the reducing the number of people required to perform various functions. In Belize fish are utilized to deal with human waste. The natural beauty of Hawaii attracts so many people to visit these islands that food production gives way to tourism. Man-made fertilizers enable countries like Mexico to become abundant food producers. And, on the other hand, Cuba's alternative political and economic approach demonstrates the need for many more farmers than the countries that emphasis the efficiency of using technologically sophisticated farming methods.

It is interesting to note that despite the fact that energy costs have spiked up once again to take the economic centre stage as they did in the 70s, Schmidt's prediction regarding the demand for fertilizer may yet come to pass. Given that the global population has reached 6 billion and continues to rise, being able to produce a sufficient food supply without fertilizer is inconceivable unless human ingenuity can develop some radically new technology or face the alternative of a dramatic decrease in human population.

BELIZE – *The Calgary Herald* – January 4, 1973

There are so many delicate connecting links in the eco-system that man has broken many of them without realizing it. If the balance of nature is upset only a small amount at a time there is a chance nature can heal the broken link by itself. A striking example of the process may be found in the City of Belize. This city, being four inches below sea level, has problems with drainage and sewage disposal. The Belizeans are a bit sensitive about their city's image in this regard.

During a recent visit there, I made an offhand remark to a businessman about the "open sewers." He immediately set me straight that there are no open sewers and that Belize has one of the cheapest and most efficient sewage disposal systems in the world. He admitted the surface drainage ditches are a problem. Water may stagnate in them for a week or so when there are no heavy rains to flush them out. Putrid dishwater, algae and urine don't smell exactly pleasant to the visitor especially in a hot climate. High tides often back into the city and leave more water in the ditches, which, incidentally, take the place of sidewalks in the narrow streets.

A team of Canadian engineers, working with a Canadian International Development Agency grant, made a study of the feasibility of providing better surface drainage. I haven't heard the results of the study or recommendations or correcting the odor situation. The need for a port from which to export lumber was the main consideration in building the city on mangrove swamps at the mouth of the Haulover River several hundred years ago. Once the city was built several canals were built to improve the drainage. People and businesses began finding it convenient to run their sewage systems into the canals. There are several public lavatories over the canals. At first the volume was not enough to cause problems. Later an increasing volume would have caused trouble except for one thing: the piranha (a species of hungry tropical fish) began to come into the canals and scavenge the human waste.

There are literally millions of these faeces-eating fish in Belize's "open sewers" today to keep the waters purified. The businessman showed me how they operate. He threw a crust of bread into the water. Instantly there was a great jumping, splashing and thrashing as sharp-toothed fish rushed for it and tore it to pieces and ate it. "If anything happened to these fish we'd be in real trouble in this hot, humid climate," he said.

To provide jobs, industry is being encouraged through tax incentives. The most recent industry to be set up in Belize, formerly British Honduras, is toilet paper manufacturing. It provides 15 new jobs, as DREE in Canada is wont to point out. To herald the inauguration of this pioneer industry, the large display window of the TAN Airlines office in the centre of town was piled high with new rose-colored tissue in one and two-ply models. (TAN is the national airline of the Honduras.) It was an unusual and eye-catching giant display for an airline office. The toilet paper was selling at bargain prices—almost as good a bargain as the raw sugar cane rum, known as "chop," at 60 cents a mickey.

Man can use his technology to secure himself against the elements but tinkering with the ecology is done at the peril of his or succeeding generations. Belizeans are also sensitive about their situation in regard to one of nature's most fearsome phenomena: hurricanes. The businessman admitted he had been through two bad ones in 1931 and 1961. Both these "blows" nearly swamped the city. The latter resulted in a decision to relocate the capital 50 miles inland at Belmopan. However, he suggested that what Canadians read about the hurricane situation on an annual basis is mostly mythology.

Annually the government issues a hurricane supplement to its information magazine giving the names of the hurricanes and precautions to follow when the national radio issues a hurricane advisory. A most pertinent precaution is that if winds are estimated to exceed 50 miles an hour as the storms approach, all coconuts should be removed from trees near buildings. The citizens are better protected today by hurricane shelters. For instance the imposing new Royal Bank of Canada building in the centre of the business district is one of the hurricane shelters. The Bank of Nova Scotia is also building a large new structure which will be another.

Many of the wooden business premises and homes are being replaced with brick or concrete structures. The banks not only protect people's money but have an interest in protecting their customers should they need security to weather a hurricane. This is a cheap—and most acceptable—solution to a vexing problem.

HAWAII – *The Calgary Herald* – February 2, 1973

At the present time there are signs of economic distress in the production of some of the crops throughout the world. Pineapple production is in trouble in Hawaii. Sugar cane may be a dying industry in the Caribbean. Cocoa, coffee and banana plantation owners in the Caribbean are having trouble. Countries which contain two-thirds of the world's population suddenly ran short of bread grains and protein last year. Meat production difficulties have been experienced in Argentina and Uruguay. Even in Alberta the government is worried about the terrific decline in dairy cattle numbers and shortage of milk in a province which used to be a big butter exporter. What's going on?

Frankly, one would have to do a bit of groping for an answer. If the world were not enlightened by an increasing percentage of literacy among the masses, it might be easy to give an off-the-cuff answer to the reason for these shortfalls. A quick appraisal might show the blacks are getting sick of feeding the whites;

the poor are getting sick of feeding the rich; the little man is getting sick of carrying the big man.

During National Farmers Union discussion of the Kraft Food Company boycott, some references were made to the "yoke" imposed on farmers by the Dole Company, the largest pineapple processor in the Hawaiian islands. Dole, a household name in Canada, is a vertically integrated operation. It was a great surprise for visiting Canadian ranchers and farmers to learn that Castle and Cooke, parent company of Dole, has made plans to drop its growing leases on Molokai and convert its Oahu operation to production of fresh fruit only. To most Canadians pineapples are Hawaii—and it is practically unthinkable the industry could be in trouble there. But that is exactly what a department of agriculture study turned up. It's being given 10 years to survive.

Lt.-Gov. George R. Ariyoshi of Hawaii called a meeting last month to decide how to deal with the industry, which the study said "is steadily declining in strength in national and world markets." The pineapple growers association officials admit small independent farmers cannot adequately produce a continuous supply at the proper time for processing. The association has given its blessing to Dole vertically integrating the industry by owning or leasing all the production acres.

An outside influence is the chief architect of the pineapple industry's decline in Hawaii. The two million tourists, who inundate the islands' 750,000 inhabitants with more money than the military spends, have created a high rate of inflation. Wages and land prices have gone to astronomical heights. The result is Dole can no longer afford to grow pineapples on land that is worth up to $300 a square foot and native Hawaiians are no longer content to work for farm wages when they see the big money available from those pale-faced tourists that are willing to pay fantastic prices for sunshine. Dole is transferring its production acreages to Okinawa and the Philippines, where costs are lower.

In the Caribbean, a number of factors effect slumps in agricultural production, including the fact of the well-heeled tourist. However, there is another factor. It is the fact the blacks have only in recent history thrown off the yoke of slave labor which operated the sugar cane and other plantations. They associate field labor with slavery and just simply won't work in the fields.

In similar frustration, George Morris of Merlin, Ontario, president of the Canadian Cattlemen's Association, gave notice beef producers are getting pretty sick of government measures to inquire into the price of their product again. He was so outraged he threatened to go before the price commission and tell it that the price may appear high, but what if the beef stopped coming? What if the cattlemen said they'd had enough of fighting for a good price for their work and simply gave up? This is a threat that could prove real.

Alberta dairy farmers are apparently taking the same point of view—although they went through an active government campaign several years ago, which forced many of the smaller dairy farmers out of production. These government campaigns made so many of them just plain discouraged at milking cows seven days a week they gave up production and now, when the government is trying to coax them back into production, they are shying away from it. As one dairyman said, "the government is penalizing me for over-production while at the same time importing product in the form of butterfat. Very discouraging."

So there you have it. It's not a complete or incisive report. It's just a couple of highlights picked up at random to indicate we're going to be paying more for our food before we pay less. Farmers are sensitive people. They'll work mightily to produce if proper respect and appreciation is given their efforts. When they feel insulted today, they stalk out of their fields in disgust or in discouragement. They are getting to be a real no-nonsense breed of people on a worldwide basis if they feel they haven't had a square deal.

MEXICO – *The Calgary Herald* – November 12, 1974

Dr. Norman Borlaug, the "father of the Green Revolution," who directs the wheat and corn improvement centre in Mexico, says an annual world investment of $8 billion is needed to meet increased fertilizer requirements. The alternative is famine for millions by the end of the decade. Earlier this year Borlaug was on a 4 1/2 month food study tour around the world. He claims the only country building enough new fertilizer plants is China. China is building eight nitrogen fertilizer complexes.

Just a few years ago, spurred by the adoption of high-yield grains in the widely heralded Green Revolution in the Third World countries, the fertilizer industry built so much capacity it was threatened with bankruptcy. This slowed down construction at a time when construction should have been maintained at a steady pace. And now, despite those new fertilizer plants in China and the new factories and mines being opened all over the world, the builders can't keep ahead of the demand. In addition, Borlaug found a shortage of chemical engineers trained for specialized fertilizer technology.

For lack of sufficient fertilizer, the Green Revolution is stalled all over the developing world. With much of the world's land suitable for agriculture already exploited, with a world population of nearly four billion growing at better than 2% a year, experts estimate increased supplies of fertilizer are essential to avert chronic famines. The hope of the Green Revolution was that it would buy time while governments acted to stabilize population growth. But now with the Green Revolution stalled and governments wasting the time gained so far without taking long-term measures, Borlaug figures it will be a fight every step of the way to provide nutritional food to the people who require it.

One of the problems about which the food producers and scientists of the world are agonizing at the present time is the deployment of fertilizer once it is manufactured. One body of thought feels it can be best used in countries like Canada and the United States where advanced technology can be put to use to grow large crops for export to those nations which are short. Another body of opinion is that fertilizer should be directed to the countries, which will grow the most food. They say fertilizer will produce yields twice as large on the nutrient-starved soils of Asia, Africa and Latin America as on the already generously fertilized croplands of the United States and Canada.

Agricultural researchers are giving new attention to how hundreds of millions of small farmers could apply smaller amounts of fertilizer to plants at just the times plants need them. As applied to the economics of India: For a lack of a pound of fertilizer costing 15 cents, there is a failure to grow 10 pounds of wheat that country must try to buy on the world market for at least $1. Although the pessimists see a Third World War shaping up over petroleum supplies, my prediction is that if a war comes, it will be a fertilizer war.

CUBA - *The Calgary Herald* - October 31, 1975

When the mail service is operating, one can expect notification every two or three weeks that John Rudiger is up to something new or different. He is a pioneer Charolais breeder who has a ranch west of Calgary. He runs a couple of production sales a year and spends most of his time in between thinking up new gimmicks or promotions to publicize them. At his last sale during Stampede Week the sale catalogue was a cut-out of his bull, Cadet Roussel. He plans another sale November 30th and he came up with a piece of mind-blowing promotion for that one. He and his wife went to Cuba on the first of October. They went on the first beef-oriented CUSO mission from Canada to teach Cuban farmers judging, selection and preparation of Charolais cattle. CUSO has sent Holstein breeders to Cuba in the past but the Rudigers' trip is the first connected to beef. They will hurry back to Calgary in time to oversee fitting their own cattle for the sale.

It is my judgment that CUSO could have selected no better representative from Canada to go to Cuba to prepare Cuban farmers to throw off the yoke of an oppressive dictator named Castro. Rudiger is a typical Western free-enterprise cattle breeder who has thrived on the democratic system that has made it possible to supply plenty of reasonably priced food for Canadians. If Cuban farmers were exposed to John Rudiger, they would have emulated his example and thrown Castro in the Caribbean. This is the only way they would be permitted to adopt Canadian agricultural production methods— methods which cannot and will not thrive under the Castro system because farmers are still considered peasants rather than entrepreneurs.

Mr. Rudiger came out of relative obscurity in Saskatchewan 17 years ago to break new ground with a new breed of beef cattle. Despite a few reverses along the way, he has had the freedom to be innovative and to attend more Charolais functions around the world than any other cattle breeder in North America. He has broken new ground in spreading the seed of top-rated Charolais around the world by not selling, but renting, his best bull to a group in the artificial insemination business. Along the way he has had time to become part owner of Herd Book International with Hayes Walker III of Kansas City. Lately he set up Rudiger Marketing Ltd. to distribute a new type of mineral feeding for cattle evolved by Talbot-Carlson Inc. of Audubon, Iowa. It is a cafeteria of ten minerals for cattle in a specially designed feeder. The cattle balance their own diets for minerals according to their individual body needs. In other words, the old cow is pretty smart and she will provide her owner a good living if given her freedom.

Some world leaders like Fidel Castro don't believe in this concept for people. However, it is symbolic and typical of John Rudiger that he would choose to distribute a free-choice product. He has had relative freedom of movement all his life. This freedom has taken him to Cuba to teach farmers there what freedom can do for agriculture. The object lesson should be patently clear to all Canadians and to CUSO: When we see Cubans given the freedom to move into Canada to teach Canadian farmers how to feed the people better than they are now being fed, then we will know Castro's system is worth investigating.

CUSO stands for Canadian University Service Overseas. Last April 18, I criticized CUSO for the methods of obtaining character references and screening candidates for placing in foreign countries. It is my contention that if an agrologist or farmer is capable of generating tax revenues for the Canadian government, such persons should be able to spend those taxes granted to foreign countries without any screening whatsoever. Unless agrologists are allowed to cut through the oppression, they might as well stay at home.

PACIFIC RIM

Western Canada was looking to the Pacific Rim countries and their well-filled treasuries as a new trading area, replacing its dependence on Eastern Canada. Schmidt had a chance to take a look at this market in its formative stages and his columns assisted in bringing public attention to the business opportunities available there, as well as some of the problems they share in common with Canadian farmers.

Regulations regarding seed patents are only now periodically being contested in the courts. However, their original adoption into legislation was hotly debated in many countries. While Schmidt argues that seed monopolization by multinationals is not sustainable because increases in commodity prices will attract more producers, his rationale may not hold in the future, particularly in the case of patented genetically altered seeds.

AUSTRALIA – The Calgary Herald – February 17, 1981

The Australian government has a plant breeders' rights bill coming up at its spring session. The Aussies call it plant variety rights. The government there does not face opposition from Pat Mooney and his mates in the Society for International Cooperation, the New Democratic Party and the National Farmers Union as does the Canadian government. It faces opposition from the Plant Diversity Protection Committee, whose umbrella organization is the Total Environment Center. The legislation is termed the most important piece of agricultural legislation ever to get onto the floor of the Australian Parliament. The legislation will allow breeders to obtain sole propriety rights over any new plant they may develop. These rights allow a breeder to control the use of a variety, to levy and collect royalties and to take civil action against any infringers. Such a property right is similar to a patent but sufficiently different to require special legislation.

Aussie informants point out the legislation has been deferred twice in the Canadian House of Commons and claim 50% of farmers here are against it. Nelson Coyle, an NDP researcher, recently told a group of University of Waterloo students Prime Minister Trudeau is against the breeders' rights bill. This has split the senior ranks of both the Liberals and Conservatives and severely wounded the pride of Agriculture Minister Eugene Whelan who introduced the bill. The bill is also opposed by Don Mazankowski, former PC transport minister, and Consumer Affairs Minister, Andre Ouellet.

The United States passed the Plant Variety Protection Act in 1970 and proposes amendments. But the amendments have brought opposition from Mooney-type protesters. In Australia, two main groups have been lobbying for the introduction of plant variety rights legislation:

1. the Industries Committee for Plant Breeders' Rights, which is said to represent and to be funded by, the principal overseas seed companies, and
2. the Australian Seed Producers Federation.

The Australian government made draft legislation available to the Seed Industries Association, Australian Seed Producers Federation, Australian Nurserymen's Association and the state departments of agriculture and various federal departments for comments by May 15, 1979. The general public did not become aware of the legislation for six months after introduction. Since then the government has consistently refused to make details of the draft available to the public.

Traditionally, seeds have always been a public natural resource. But increasingly they have moved into private hands. In past times public breeding has ensured the development of varieties based on the needs of growers with free exchange of genetic materials taking place. The opposition to the legislation is concerned that private companies or multinationals will take over seeds and this could result in adverse economic, biological and social effects.

Ultimately the scenario goes, 17 companies could control the world food supply, just as 17 companies control the world's supply of grain. In other words the Australian and Canadian governments would be helping to create a seed OPEC. The multinationals are buying up small seed companies all over the place. Well, this might work for oil. But it will never work for agricultural commodities.

The Australian, U.S., Argentinean and Canadian governments have been trying to get a new wheat OPEC going. They have failed—and they will continue to fail, because once the price of wheat gets high, it becomes attractive for everybody to grow and surpluses result. The same applies to seeds. If the price of beet seeds goes so high, I will start growing my own in my vegetable garden and trading them among my neighbors just like the pioneers did when they had no money to buy them. It's just that simple.

PHILIPPINES - *The Calgary Herald* - January 4, 1983

In a recent column some views on plant breeders' rights were presented. New Zealand plant breeders say farmers have benefited from legislation passed in 1973 allowing them to patent new varieties and collect royalties. The opponents of plant breeders' rights in Canada have tried to convince us Third World countries would suffer as a multinational agribusiness mafia would spring up and get its hands on many hybrid seeds and their farmers would be denied many useful varieties.

Let's take a look at the other side of this picture in a country regarded as being in the Third World, the Philippines. A great deal of foreign aid has gone in there, including Canadian dollars. Following the development of hybrid rice varieties, the Philippines has now become an exporting country and may add corn to its list of export commodities. Corn? At present the Philippines imports about 235,000 tonnes of corn worth about $40 million from the United States, Thailand and Australia every year. Corn is one of the diet staples for many of its 48 million people. Corn is also used to feed an expanding poultry and livestock industry. Imports have doubled in five years. How can it become an exporter in four years, as is predicted by Agriculture Minister Artruro Tanco?

The answer is bringing the San Miguel Corp. into the department of agriculture's maisagana program to increase corn production. San Miguel is a giant beer and food conglomerate. It is big in the manufacture of poultry and livestock feeds; that makes it the largest corn consumer in the country—120,000 tonnes a year. San Miguel was in corn production in the early 1960s until rice and corn were nationalized. The legislation was repealed in 1975 when the government saw the error of its ways as production of both fell drastically. The company went back into production and started an experimental farm at Calaun, 60 km. south of Manila, to develop hybrid corn varieties with the assistance of the International Centre for the Management of Corn and Wheat in Mexico. It also has a seed-processing factory and recently opened SMC Seed Research Centre.

After five years of research costing $6 million, its scientists have come up with two hybrid varieties which are regarded as "wonder corn" in the Philippines. They will withstand downy mildew and corn borer, two dreaded pests that are the scourge of Filipino corn growers. The two new hybrids will yield 3.6 tons per hectare, which is four times the yield of most corn now grown. The company has another variety in the hopper which will go eight tons. With a quadruple output, Tanco expects acreage can be cut to 300,000 from 3.1 million and this could put the Philippines in the export business in four years.

Japan and Taiwan import $1.5 billion worth of corn annually and Tanco is positive his country could capture 10% of that market. This could change a $40-million foreign exchange deficit to a $150-million earning. In copying the rice production program, the Philippines government launched the $12-million maisagama program last December. Its first attempt in 1974 was a failure because it didn't have the plant-breeding expertise which San Miguel later introduced. Under this program, good farmers are selected and money is loaned to them to get into hybrid seed, fertilizer and pesticides but the higher production more than covers these added costs.

Kenya also went this route for increasing its production of corn, using two private companies to develop hybrids and move its farmers into them. It reversed to an exporter from importer, although the government ran into trouble by exporting too much several years ago. A couple of these success stories have given some of the Third World countries new confidence and braggadocio.

Modern Asia Magazine, published in Hong Kong, boldly predicts 1,000 new multinational companies based in Third World countries are going to give those in Western countries a good run for their money in agriculture and industry. Such companies in Korea, India, Taiwan and Brazil are successfully competing for world markets and in the next decade could provide a strategic challenge to the United States.

This is what plant breeders' rights critics in Canada have omitted to tell the people. If the U.S. can do it successfully, there's nothing to stop Third World countries.

NEW ZEALAND – The Calgary Herald – November 26, 1982

Agriculture Canada had planned to start using the Hennessy electronic grading probe for hogs late this year. However, the inventor, Dr. Brian Hennessy told me a delay may be expected. I was one of 140 international farm writers who visited Dr. Hennessy, an animal scientist at Ruakura Animal Research Station at Hamilton, New Zealand, November 3, 1982. He gave us the first public viewing of the probe he and his associates invented. It looks like a Star Wars zap gun on a coaxial cable—except the working end of the muzzle is a razor sharp triangular knife blade. Fitted into the base of the blade is a glass-enclosed electronic eye. The carcass is zapped by inserting the probe between the 12th and 13th ribs adjacent to the centre of the eye muscle longissimus dorsi to measure the subcutaneous cover fat thickness.

In Canada, Agriculture Department graders use the measurement of the depth of this fat to determine what producers are paid for their hogs. Until last March federal graders used two physical cuts—between 12/13 and 9/10 ribs—to calculate the grade. In March, they went to one measurement, as the government had decreed the optimum site for the measurement is between the 12/13 ribs. There were some charges by Western producers that the one-measurement grading had resulted in lower payments to them. Brian Glanfield of Edmonton, who heads the livestock grading standards office for the department in Alberta, said criticism has now abated. However, Dr. Howard Fredeen, livestock geneticist at Lacombe research station, alleges the one-measurement grading had not been subject to proper research.

Producers were told the electronic probe would replace this system before the first of the year after a series of tests at a packing plant in Toronto. However, the instrument being used broke, thus delaying the testing. In the meantime, the Canadian government had ordered three more of the $10,000 instruments for tests in other parts of the country. Up to November 3rd Hennessy had not been able to forward them due to delay in receiving key parts which he had to import. The inventor said the probe is made to take one measurement on 1,100 carcasses per hour on a moving kill-floor chain or two measurements at a rate of 500 per hour. The blade goes into a hot carcass easier. But the demonstration at Ruakura showed it took a strong man with big muscles to push it into a cold carcass. The cold carcass tends to swing away and make it difficult to pierce.

The critical part is the electronic eye behind the blade. That's what broke in the tests in Toronto. It may have to be modified so that the muscular person won't break it, particularly if it strikes a bone on the way in. However, once these problems are overcome it will be valuable to the meat industry as it will eliminate human error in grading, said Glandfield. It may be necessary to make two measurements as Fredeen said that although the single measurement has been justified as part of a gradual move toward electronic grading "increased sophistication of the measuring device is no guarantee of improved carcass evaluation."

INDONESIA - *The Regional* - February 15, 1999

Over the years the Canadian International Development Agency (CIDA) has received a great deal of criticism for its unwise and costly programs overseas. However, one of its costliest, long running projects received little, if any, praise or publicity despite the fact 30 scientists (mostly from Alberta) were employed. These scientists worked for two years from 1974 under contract to CIDA writing the massive East Indonesia Regional Development Study. Some of the scientists lived for two years in Bali compiling the 14-volume technical report plus a compendium of project proposals and a general strategy design. Others spent shorter tours.

The bilingual report (Indonesian and English) weighs 27 pounds. I am the proud owner of a copy, courtesy of Cy McAndrews, who was the program leader. He was then Alberta deputy minister of agriculture; he is now retired in Edmonton. McAndrews was rather pleased to have a farm writer look through his career magnum opus, indicating that I was the only one to have shown an interest. A copy of the report has been sequestered in the Alberta Department of Agriculture, Food and Rural Development library in Edmonton for over 20 years. The report has not suffered the same fate among the professional agriculturists in Indonesia. It is still being used in some provinces of East Indonesia as a blueprint for development and will continue to be used for several more years. Each Canadian scientist was assigned an Indonesian counterpart to carry out the work after Canadian scientists silently left for home.

At the time the work was produced, Canadian interest in Indonesia was of low intensity, especially of that tropical paradise's agricultural development. It was not until the Bre-X gold mine fiasco that Canadians were sent running for their maps that interest in that South Pacific archipelago became pronounced.

Pat McFayden, who was working as an extensionist with McAndrews and John Calpas at the time they were assigned to the project, said she went all over town trying to find a map of Indonesia. She finally located one in a *National Geographic Magazine* in a second-hand bookstore in Edmonton.

Although I was covering the farm beat for the Calgary Herald at the time, I had heard not a peep about the project, either through contacts in CIDA or the agriculture department. I'm not sure whether the semi secrecy was planned or accidental. I knew on a first-name basis many of the personnel involved such as Lloyd Rasmussen, Orlon Bratvold, Aubrey Sherman, Mel Lerohl, Dr. J.J. Richter, Wilf Cody, Dr. Harold Love, S. Mahadeva and Dr. Helen Abell.

Why did CIDA pick this little known country, whose population was then 130 million, to invest money in agricultural research? The chief reason was that before the Suharto regime threw out the Communist regime of the "bad guy," Sukarno, in 1964-65, the Soviet Union was pouring a lot of foreign aid into Indonesia. It then became incumbent upon the West to take up the task and CIDA was the nominal agency for Canada's contribution. This was even though Suharto developed into a new "bad guy" and got booted out of power in 1998.

By Canadian standards, Suharto was a "bad guy" from the first. He rounded up 10,000 political prisoners in 1965 and exiled them to the bare bones island of Buru without any agricultural implements. Not all of them survived this outrage.

"The Indonesian administration wouldn't let us go there to make an assessment of this New Order experiment," said McAndrews. Why did this obscene forced migration not rate headlines in Canadian papers? Not even the Australian-based magazine, Inside Indonesia, which details present-day atrocities in East Timor, ever makes a mention of Buru today.

GLOBAL AGRICULTURE

Schmidt discovered that ministers of agriculture and farm leaders of various countries write one speech and pass it around to each other. The result is they have all abandoned as futile such seminars as the Federal-Provincial Agricultural Outlook Conference that Canada used to hold every December in Ottawa. At the time, it was considered quite startling when Schmidt suggested the wrong people were invited. Even more astonishing, even today, is the realization that the people who have the greatest influence on Canadian farm prices aren't necessarily Canadians. World trade in agricultural commodities is a funny business. Contrary to popular misconceptions about global food supplies, he quotes Jim Romahn as having pointed out that surplus agricultural production causes more difficulties than shortages.

GLOBAL MARKET - *The Calgary Herald* - January 26, 1977

Recent comments by farm writers Henry Heald of Ottawa and Jim Rusk of Toronto about the Federal-Provincial Agricultural Outlook Conference held every December in Ottawa are that its format should be revamped. If this ever comes about, I offer the sponsors free some of my thinking on necessary changes.

I had a small exchange of correspondence with Heald a couple of years ago when he was working for the information division of the Canada department of agriculture sending out invitations for the conference. He was an enthusiastic advocate then—but now that he is not with the department any more he has done a switch and thinks the conference should be retired with full honours. His alternative is to develop a new format. I'm glad he suggested a new format, rather than abandonment. There are a great many rich farmers who would hate to see reports from the government agricultural economists cut out altogether. They read the outlook and recommendations avidly—then do exactly opposite. That's why they are rich.

A copy of my correspondence with Heald is probably stored away in the "top secret" archives in Ottawa. If I remember correctly I suggested they were inviting the wrong people to the Outlook Cocktail Party—an important wheel-greaser which Jim Rusk dared to reveal. Two years ago the government had to postpone the outlook as prices were fluctuating so wildly nobody would venture to commit himself. The persons who have the greatest influence on Canadian agriculture aren't necessarily Canadians. Just think that over.

Over the last few years the president of the United States and the secretary of agriculture have had more influence on Canadian agriculture than anyone in Canada. Decisions made in the Kremlin in Moscow have had a decided effect on Canadian agriculture. So have that little group which meets in a Chinese laundry or somewhere in Peking. The Australian Meat Board has damn near castrated the Canadian beef industry. And for a couple of years some fishermen off Peru drastically affected the feed business. Those people driving Limousins in France changed the course of Canadian cattle breeding for a while.

Rusk pulls back the curtain to reveal the sponsors of the Outlook Cocktail Party are the agricultural attaches of the foreign embassies with large expense accounts for entertainment. They would be persons like Ludvig Ingersley Madsen, agricultural counsellor to the Royal Dutch embassy. It is easy to speculate that the reason for this sponsorship is to put the Canadian participants in a euphoric state to ignore increased foreign imports of agricultural commodities from the party hosts.

There have been hefty increases in imports of agricultural commodities in the two-year period ending in 1975. To really find out what is likely to happen to Canadian agriculture, it would only be necessary to pull a switch on these foreign attaches; have the Canada department of agriculture get them boozed up and get their tongues wagging. Better still, set up protocols to have them invite some of their key decision makers back home to the Outlook Cocktail Party. As the grape started flowing the Canadian hosts would surreptitiously switch to sauerkraut juice or apple cider specially imported from Waterloo County. No doubt such a format would provide in its most eclectic form the political savvy Henry Heald seeks in a new outlook conference format.

OPEC – *The Calgary Herald* – September 12, 1977

Alberta Premier Peter Lougheed and John Channon, chairman of the Alberta Grain Commission, should pay another visit to the Shah of Iran. They should take President Jimmie Carter of the United States with them. They should walk up and rap on the palace door in Teheran and remind the shah about some of the facts of life about which he was holding forth four years ago. He said the main reason the OPEC nations were raising their oil prices was that wheat prices had gone up to around $5 a bushel and OPEC was entitled to raise oil to offset the food price.

Lougheed and Channon need to impress upon the Iranian ruler that wheat has dropped to about half that price now—and does he plan to cut his oil price in half? Canadian and American farmers can afford to take less for their wheat if their input costs are lower. Petroleum products and petroleum-based products are important inputs and they have skyrocketed. It's a vicious circle which only the shah and Lougheed can do something about.

Carter will find out, to his regret, that he may have over-reacted in ordering a cut of 20% in wheat acreage and locking much of the "surplus" production into reserves. This move will have the effect of forcing up wheat prices again—only the next time the prices will shoot up to $10 a bushel. If this happens gas prices will be forced so high by OPEC that Lougheed and Channon won't be able to afford enough gas to jet into Teheran. And if they do they will probably be greeted by the shah with a gun.

A great many people laughed when Jim Romahn, farm writer for the *Kitchener-Waterloo Record*, said two years ago surplus agricultural production would cause more difficulties than shortages. He was right. The world wheat trade is a funny business. It would surprise many persons to know at the present time India has some 23 million tons in storage surplus to its immediate needs. Of this amount, 10 million tons are in danger of rapid deterioration because the nation doesn't have proper storage for this amount. Much of it is bagged and sitting on unused airport tarmacs, open to the ravages of rats and rain. India has seen the wettest monsoon season in years.

There is currently enough wheat in India that the government feels it can ship back two million tons it "borrowed" from Russia several years ago. People "in the know" on the Indian wheat-marketing scene say that even when that country was a big purchaser of wheat—due to bad harvests—huge quantities of it were being transhipped to Russia and other countries. That would be the answer to the puzzle about India suing several American-based grain companies for several hundred million dollars for under-grades and high dockage for wheat delivered over a period of 12 years. It is within the bounds of speculation that the ultimate recipient has egged on India to inaugurate the lawsuit. Anyone who has ever been through the bazaars and seen their own wheat and flour dealers sorting out the wheat grain by grain knows it is being done for a good reason: to get the dockage and under-grades out.

The Carter move will take 22-1/2 million acres out of production. That's almost as much as the total Canadian wheat acreage. Although there may be pressure from the U.S. for Canada to follow suit and reduce acreage next year, there will be great political resistance to this move. It is doubtful if Wheat

Board Minister, Otto Lang, would even suggest it to growers as too many remember his infamous LIFT (Lower Inventory For Tomorrow) program of 1970. The only hope the government has of achieving anything like this in the West would be to ask Jack Horner to do it—or cut the initial payment as should have been done this year to achieve the same result painlessly. In the meantime, everyone is keeping an eye on Argentina where a new government policy has made for an increase in production of all grains—corn, wheat, sorghum and coarse grains—of 35% last year over 1975.

AGRICULTURE SPEECHES
– *The Calgary Herald* – August 26, 1983

It almost seems as if the ministers of agriculture and farm leaders of the various countries have one of their number write a speech then pass it around among themselves. During a luncheon for the press corps covering the Palermo Show in Buenos Aires August 18, Horacio Gutierrez, president of the Argentina Rural Society, delivered a speech whose contents have been voiced many times previously.

As it was an international meeting, delegates of many nations came, including a group of 20 Western Canadian cattle breeders and bull semen exporters, led by Jaime (pronounced I'm-ee) Ellehoj of Semen Exporters Canada Ltd. of Balzac. I was fortunate enough to be included. The visitors gave Gutierrez a chance to have critical exchanges of opinion on agricultural production. He firmly believed ministers of agriculture are very important persons as they are responsible for feeding millions of people as they draw up plans for grain production. "I talked to the ministers of agriculture for China, India and the United States," he said. "They are responsible for feeding half of the world's population, two billion."

Gutierrez added, "the Chinese minister told me that Argentine farming methods are extremely efficient and that although China was a purchaser of food from Argentina the possibilities of increased trade are very much greater since he had to make plans to feed a billion people—about 23% of the world's population." Such statements make Argentine producers realize they have added responsibilities if they are to produce this much-needed food in the future. Gutierrez made a complaint familiar to many Canadian farmers. He complained about government ad hoc agriculture policies that failed to give farmers a clear indication of national production intentions at a profit.

Last Saturday night at a dinner at the Danish Club in the city of Nechochea, Ing. Victor Hugo Santirso, the Argentine minister of agriculture, paid the Canadians a special surprise visit that lasted two hours. (Ing. denotes he is a professional engineer.) Ellehoj is a native of the Nichochea area and a long-time friend of the minister. Santirso said that to meet the needs of the world's hungry, the Argentine government plans to increase grain production (mostly wheat) to 60 million tons annually by 1990 from 40 million today. Of the total, 45 million tons would be available for export. To accomplish this, the national cattle herd would be reduced. In 1978 it was 61 million. It has since been reduced to 53 million, although drought is partly responsible. As grain production has increased, sheep production has also gone down to 50 million today from 70 million in the late 1970s.

But having decided to prime up the nation's agriculture for expansion, Argentina faces the nagging problem of the European Economic Community (EEC), the world's most powerful and protective trading bloc. This bloc has a great deal to answer for, not only in the eyes of Argentina but Canada, Australia, New Zealand and the United States. It stands accused of preventing the low-cost exporting nations from moving farm commodities into the EEC member nations by the use of tariff protection amounting to $14 billion a year and of stealing their traditional markets.

Santirso still has the problem of inducing farmers to increase production, despite a 25% export tax that hangs around their necks. It has been an off-again on-again tax that has been a definite deterrent to production when it has been applied. Can he increase production if it remains? "As a farmer, I believe it should be repealed," he told me. "But I can't convince the revenue minister it should be." There is a ministerial struggle going on, but the revenue minister has to raise money somehow so Argentina will be able to pay off its crushing $40-billion foreign debt.

GLOBAL BOUNTY – *The Calgary Herald* – August 13, 1986

The Australian consul general in Vancouver advises that the agribusiness consulting firm, ACIL Australian Pty. Ltd, with 23 years in business, is available for hire anywhere in the world to increase livestock and food crop production. While it is laudable for consulting firms to line their pockets by doing this kind of humanitarian work, the end results of providing bounteous supplies of relatively cheap food for human populations are often discouraging. There is endless carping and complaining about food quality when copious supplies are available.

Starting with the strontium-90 scare in the 1960s, if everyone had quit eating foods that had been alleged to be contaminated by chemical residue, we would all be on a diet of bananas. They are about the only food I can recall which has never been condemned as injurious to human health. What does it profit a nation to have the means of production of huge quantities of good food then fritter it away through fads? Because of the topnotch animal breeders and first-class government veterinarians, Canada has achieved a worldwide reputation for disease-free breeding stock. Yet the same veterinary service, which has achieved this reputation, tends to be secretive about the reports it writes on contamination in Canadian poultry, meat and fish processing plants.

Occasionally word gets out on the street about contaminated product reaching stores but it's hard to assess whether those reports are alarmist or whether we are too fastidious about sanitation. At the present time, the Canada Department of Agriculture is attempting to keep under wraps the reasons for contaminated meat from specific plants reaching retail outlets. News of these incidents reached the press through leaked documents. Word about marketing boards and health authorities cracking down on farmers who allow drug and antibiotic residue to go to market in pork or milk create some excitement—but not too much. In the case of milk, offenders in some provinces are being penalized so heavily they have nearly been put out of business.

One of the issues discussed at the recent Canadian Institute of Food Science and Technology annual conference in Calgary was the increasing evidence of food-borne disease in Canada. It's not only scary but costly and may even be beyond the ability of the inspection system to control, it is claimed by Dr.

Ewen C.D. Todd, Ottawa, who is with the bureau of microbial hazards of the health protection branch of the National Department of Health and Welfare. He had figures to show that the reported incidence of salmonella infecting human beings in Canada was seven times that of the U.S. in proportion to population. The total cost to the Canadian economy for this and other food-borne disease is $10 billion a year in Canada, he claimed. For instance, 599 members of the Ontario Provincial Police were hit with food poisoning during the Pope's visit— and the Ontario Police Association is looking for someone to sue to cover their losses, he reported.

As bad as it sounds, Todd believes many cases of salmonella go unreported either because of the mildness of the illness or ignorance of the cause of an illness that comes and passes quickly. From a layman's point of view, it is hard to assess the significance of Todd's remarks. I find difficulty in judging whether his description of food-borne disease is normal or whether bugs and bacteria are running away with food after it leaves the farm. Do Canadians need lessons in simple food hygiene? Is he using scare tactics to extract more money out of the government for inspection and control? Would lack of such money turn us into a have-not country from a "have" nation?

Another of the issues about North America's abundance of food, which has baffled the food scientists, is the fads which have switched consumers away from good, protein-rich, nutritious foods. As Percy Gitelman, president of the Alberta Industrial Mustard Co. Inc. of Lethbridge, put it: "The dramatic drop in consumption of beef is a fact which was not—and could not—have been predicted 10 years ago. And just as the drop in red meat consumption was unpredictable, so was the dramatic success of chicken, which is a finger food."

Members of the medical profession, posing as nutritionists, have been fingered by the cattlemen as the chief culprits in this huge upset in agricultural production. The question is: Would the cattlemen see a reversal of their fortunes if the doctors in Ontario had stayed on strike forever and ceased handing out this questionable advice?

World wide, agriculture has been and continues to be humanity's most important industry. Over the centuries literally thousands of talented people have employed their creativity to enhance nature's bounty. While the ability to produce food in large abundance is remarkably successful, the distribution of food remains shamefully unequal. Nonetheless, releasing humans from backbreaking toil continues to be a mixed blessing, particularly in regions where technology replaces human labor too quickly. In areas where people are not equipped to participate in economic activities beyond subsistence farming, introducing modern farming technology may produce much more food more efficiently, but leaves many people languishing without recourse to gainful employment and little access to the bounty.

Despite the purported efforts of individual countries to pool information through various international organizations, many problems persist in sorting out the creditability of the information and disseminating it in an ethical manner. Questions regarding whose purposes, at bottom, are being served go largely unanswered. However, there are some columnists who don't simply parrot the obvious. Like Schmidt, they attempt to piece together the bits of really meaningful information, in an effort to present a more unified picture of the whole, to enable a better understanding of the events and issues that have such a huge impact on our lives.

The work of all journalists working on foreign assignments is arguably more diffi-

cult than reporting on domestic situations. The job of making sense of the world food production and international trade is extraordinarily intricate and convoluted; it requires someone with the breadth of knowledge and experience of a John Schmidt. Particularly if they have the ability to convey the information in a reader-friendly, understandable manner.

CHAPTER FIVE

Farmers Frank

HUMOR BEST MEDICINE

Getting up early in the morning enables Schmidt to enjoy a couple of good naps prior to lunch, but, come midnight, he is ready to debate with anyone about almost anything. Between naps and heated debates, there is really nothing he likes better than to fit in as many smiles and chuckles as he can. He strenuously maintains that his status as a "senile citizen" qualifies him to prescribe laughter as the elixir for a long and healthy life.

REVERSING RESOLUTIONS
– *The Calgary Herald* – January 3, 1977

It's in one year and out the other. No, Thaddeus, the Western Stock Growers Association didn't invent New Year's Eve—a night of revelry which represents the triumph of optimism over experience. Members of WSGA tell me they are never again going to make New Year's resolutions. Things have been so bad in the last year in their business that no matter what right, good and beneficial things they resolved to do, all went sour. So taking a leaf from their tally book, I have slung a different slant on my 1977 New Year's resolutions. I hereby solemnly resolve that in 1977 I will:

1. Start smoking again. (Some of my tobacco-growing friends around Drumbo and Bright, Ontario, will go to the poorhouse if I don't. Or worse still, they'll have to go into corn growing and this will reduce the market further still for Western feed grains.)

2. Write nothing snarky about marketing boards. (Some of my friends are employed by marketing boards. It would be a disaster if any boards closed down and they had to go on the road selling or marketing or, horrors, back to the farm.)

3. Never look for silver linings. (Cattle feeders have turned to feeding lambs. They made all the tax write-offs they can afford in cattle feeding—so they went looking for a silver lining in a business largely ruined by Australian

and New Zealand imports. That's just like buying a blue serge suit for diagnosing the dandruff quotient.)

4. Learn some new cuss words. (There are plenty of places to use them in reference to government policy—and perhaps rent a few to Alberta dairy farmers who may find they owe the milk processors money before the end of the dairy year for overproduction.)

5. Start writing cornball nostalgia and make young punks like it. (I'm sure things will come full cycle and everyone who wears blue jeans will have to start working again.)

6. Never catch up on overdue correspondence. (If people think my opinions are worth more than a 12-cent stamp after March 1, 1977 they can phone me at equally inflated phone rates.)

7. Refuse to be unbiased and tolerant. (Unless pinkos and reds can be induced to drop their "healthy biases" towards whites, all whites are going to turn onto the biggest apologists on this earth. With the job they are doing in North American agriculture, frankly I can't see the whites have a thing to apologize for.)

8. Refuse to accept criticism. (Without hitting back, that is. There are too many malcontents, perverts, minority groups, political activists, ego-tripping females, idiots, punks, jailbirds, racists and consumer groups who think that if a columnist won't support their cause without question the newspapers should come under government control. That's like buying a saddle because two people say you have ears like a donkey.)

9. Go back to my old sylph at an eighth of a ton. (Dieting is nonsense when, with the eating of a few garlic rings, I could help eat more hog producers into prosperity and with an extra bottle of rum a week drink the Lougheed government into solvency.)

10. Always draw to an inside straight. (The economists and the government would advise farmers they are crazy for doing that all the time. However, the ones that do are the only ones making any money these days.)

11. Quit trying to learn something new about something different every day. (So what good does it do if I learn the price of 50 different things are up every day when what I really want to know is why civil servants can get raises over the AIB guidelines any time they apply? What I've also got to do is get one commodity price—like gasoline—rolled back to New Year's 1976 without the collective government and petroleum industry threatening to commit suicide.)

12. My final resolution is that may the best day you ever had be the worst day you have yet to come. Happy New Year!

Technological change does not stop with food production. It impacts all aspects of agriculture including farm management. Modernity frees farmers from spending their time inspecting plant leaves to devote themselves more to turning over the kind found in the many multi-copy documents they are now compelled to fill out.

BUT FARMERS WORK FOR NOTHING
— *The Calgary Herald* - February 3, 1977

After Dr. J.G. Carter of Okotoks refused to fill in the January 1, 1977 farm census form (*Agriculture Alberta*, January14), I got to thinking what awesome havoc farmers across Canada could cause if they all went on strike. No, I don't mean a product strike but a paperwork strike. What an immense pyrotechnic display they would create across Canada if they all touched a match to all the forms the governments expect them to take time off to fill in. If at the appointed hour of 2 p.m. February 30 they all scratched matches on their breeches, they could act at one giant stroke to make time that would be better spent in the curling rink or an extra week in Hawaii.

Last year, being census year, the farmers had to fill in—in addition to the long household census form—the census form for farming, a quarterly pig survey report, a confidential farm survey and others too numerous to mention. There are highly paid auditors, bookkeepers, inspectors, helpful souls, agents, busybodies, statisticians, persons invested with authority to enter premises, fieldmen, policemen and just plain snoops from between five to seven government departments inspecting books, each demanding the farmers do reams of paperwork for nothing.

There is a suspicion much of the information is of no use and is never checked by anyone in government after the farmers go to the trouble of compiling it unless there is a sale for it in the moonlight. Farmers face a penalty of 30 days or $100 for failure to gather up the figures. Many people are wondering what is there left for the government to ask about or pry into.

Only a small amount of imagination would conjure up the impact a paperwork strike would cause on the national economy. The greatest impact would be the vacuum created. Thousands of persons, an estimated 200,000, would become instantly unemployed—starting with the loggers who cut down millions of board feet of lumber a year to supply the paper or which thousands of printers who print up the questions and multiple choice answers. Should all paperwork be struck out the day after tomorrow, so many trees would grow in this country that Saskatoon might revert to Druidism. With no work for them to do and no possible work for which they could be retrained, thousands of statisticians and economists would commit suicide. More would jump out windows than during the Wall St. Crash of 1929. The post office would have to lay off at least 2,000 people—but this wouldn't matter as they'd be able to work full time at the jobs at which they are now moonlighting.

The travel agents would be swamped with business from farmers who had an extra week or two of spare time with no forms to fill in. The benefit accruing from the strike would be that farmers would start to make money. Since the government wouldn't know anything about their business, it wouldn't be able to devise any of the silly-ass programs, which have caused laughter, tears, hardship, curses and disruption in agriculture since DBS changed its name to Statistics Canada. Statistics Canada's rolled-umbrella brigade has forced thousands of farmers to put their wives and chartered accountants to work to supplant cowboy arithmetic formerly done on the back of a shingle with cost accounting. Those who have forced this nonsense on the farming populace deserve to be the victims of a paperwork strike.

In addition to pointing out the different farm management techniques that farmers had to cultivate over time, Schmidt provides a fascinating comparison of farming conditions and methods between Russia and Canada.

COMPARING FARM LIFESTYLES
– *The Calgary Herald* – May 10, 1978

Craig Whitney is a *New York Times* reporter. He was one of a group of newspaper correspondents taken by the Russian government to the 79,000-acre Chik state farm in Siberia last February. Purpose of the visit was to explain why weather—not the Communist party's stolid state farm system—is responsible for most of the shortfalls in the USSR grain harvest. To an Alberta farm writer *The Times* writer could well have been describing conditions on a visit to the farming area around Czar, Alberta, in mid-winter.

Listen:

Craig Whitney: Winter buries Chik state farm—two days' train journey from Moscow—under a sea of snow whose undulating swells stretch endlessly across the Siberian plain. The snow started falling in October. It will not melt until April. The temperature stood at a "quite normal" 20 below Celsius.

Czar, Alberta: This agricultural community is two days' journey by train from Ottawa. It is on the fringe of the undulating Palliser Triangle. Snow comes in October and may go away in April. In fact, snow falls here in every month in various years. If they're lucky the temperature will be up to 20 below Celsius in February.

Craig Whitney: Aleksandr Kuznetsoy, deputy director of the Chik farm, describes the region as a "marginal agricultural zone." "We do not have enough rainfall," he explained. "Only 12 to 14 inches a year come and most of that in August and September in the middle of harvest." Farmers here have only 20 days to bring in the crops before the first frost comes. If rain falls during those 20 days, the harvest is spoiled.

Czar, Alberta: It is semi-arid here. It's absolutely drenched if rainfall is more than 16 inches annually. The annual rainfall pattern is such that if rain doesn't get the ripe crop, frost will get it September 1st or 2nd. Otherwise it can be taken off in a couple of weeks, provided farmers are not held up for parts by the farm machinery dealers.

Craig Whitney: "Last year was a bad year," Kuznetsov said. "We had a drought." The flat fields produced 19 bushels of spring wheat per acre though the farm's five-year plan had foreseen 27—which would have brought Chik up to productivity of the grain belt in North Dakota. About 12% of the nation's crop here is grown in Siberia.

Czar, Alberta: Drought conditions existed here last year. Fortunately rains came along to bring yields higher than anticipated initially, once again foiling the doomsayers. An inch of rain at the right time will give this area a good crop. The last man to map out a five-year plan for this area was Capt. John Palliser in 1859. He said the Palliser Triangle was a dust bowl. However, since the agrologists learned how to use the land, crops have ranged over 27 bushels an acre.

Craig Whitney: A lot of what is wrong is simply the weather.

Czar, Alberta: Albertans talk about the weather a great deal. They even try to modify it. However, they have developed new crops to harvest. The chief new one is crop insurance. It will make a $39-million pay-out in Alberta this year.

Craig Whitney: Long before Communism and collective farming came to these latitudes Russian farmers had to try to get all their field work done in only four to six months compared to eight or nine for the US Midwest. The short working season means, among other things, that during the rest of the year the state farm employees have little to do.

Czar, Alberta: This is true in this area also. However, if it were not so, attendance at farm meetings would suffer badly. And there has to be time to take the mid-winter break in Hawaii.

Craig Whitney: More households now have television, beamed from Moscow by satellite, Kuznetsov said. There are 60 autos for private use by farmhands and there is a waiting list of 50 for new ones, with deliveries coming at a rate of five or six a year. Aleksei Stupnik with 25 years' service here, who was hitching a horse to a stone boat, said he holidays in Novosibirsk, 30 miles away on the Trans-Siberian Railway, when he tires of the rural life. "We get permission to go about six times a year," he added, "and the rest of the time, well, we just stay at home and watch television."

Czar, Alberta: There is a great deal of TV watched here, most of it beamed in from Hollywood. Tiring of this, local farmers jump into pickup truck and roar into town to curl six nights a week. Or they receive permission from their wives to break the long winter isolation by frequent trips to the local bar.

Craig Whitney: Keeping the peasants on the land has been a problem since Czarist times. About 8,000 a year flee the isolation of farm villages in western Siberia and move into Novosibirsk looking for urban comforts and industrial jobs. The average wage for farm workers, according to their supervisors, is $250 a month. Industrial workers in Novosibirsk get about the same, but their apartments are more comfortable and they get almost all of the meat produced on the state farm.

Czar, Alberta: Most of the farm kids here head for Calgary or Edmonton; even the bright lights of Lloydminster. This is the manpower pool that greases the oil industry. Wages are not $250 a month but start at about $250 a week in the oil patch.

Craig Whitney: The managers in the main administration building work under a sketch of Leonid I. Brezhnev, and in the next room Marx and Engels look down from the walls. Outside stands a two-seat outhouse, the only toilet facility.

Czar, Alberta: Beside photos of Dick Damron, those of Eugene Whelan or Otto Lang are hung. But not in the outhouse. A two-holer these days won't get past the county board of health or the development appeal board.

Craig Whitney: In the state store across the street there is no meat and a white-gowned saleswoman ladles fresh milk into jars from 50-gallon galvanized cans. Kuznetsov is indignant when asked whether it has been pasteurized. "Don't try to push us down further than we really are," he said, as if pure raw milk from the farm was a sign of backwardness.

Czar, Alberta: Although pasteurization is not compulsory in Alberta, nobody drinks raw milk any more since the most recent brucellosis flare-up in dairy cattle. Milk from the city is available in all rural stores in two-litre cartons. Each farm home has a deep-freeze full of prime beef.

Schmidt's attitude toward women swings quite radically from one article to another. One moment he makes rather disparaging remarks about them as when he employs his alter ego, Thaddeus McMurphyvisk, to say: "Most women are crying or whining about something most of the time." Then, Schmidt turns right around to devote a column to promoting a good relationship between farm husbands and wives.

THANKS FOR HELPING
– *The Calgary Herald* – March 31, 1979

This is a well-travelled homily sent in by an anonymous reader who feels our readers would enjoy it. I cannot quarrel with that. The reader, presumably female, said she came across it in a Peace River weekly newspaper. It was in the column of Virginia Nell, an assistant home economist. It was written by Mrs. George Melber of St. Lawrence, South Dakota, and published in a paper somewhere in the United States. I hope somebody clips it and sends it to the *Alberta Consumer* magazine or the *National Geographic* magazine.

My anonymous friend says she hopes it won't be insulting to some readers. She probably fears some feminist bully with big, laced boots is looking over her shoulder. Any who are so inclined are so forewarned and should turn to Garnet T. Page's column or the Chicago options. Here it is:

Ten Commandments for the Farm Wife:

1. Thou shalt not sort cattle with your hands in your pockets. (Husbands and cows don't like that.)
2. Thou shalt cook meals which can be served 30 minutes early or two hours late.
3. Thou shalt learn to keep farm records. ("When did we turn out the bull?")
4. Thou shalt love the smell of new mown hay, freshly plowed earth, sweet smelling silage and the stinging sensation of ammonia in the sheep barn.
5. Thou shalt be inspired to see the sun rise and relieved to see it set.
6. Thou shalt learn to open gates, close gates and guard gates.
7. Thou shalt thrill at the birth of a new calf, and the sight of a bright new tractor.
8. Thou shalt live closer to God with faith to exceed many city dwellers.
9. Thou shalt cherish meals together, long nights of waiting for the vet to arrive and decisions about plowing up the winter wheat.
10. Thou shalt be exalted at the brotherly hand on your shoulder, the kiss on your forehead and these three precious words: "Thanks for helping."

Words and data are Schmidt's stock in trade so to sharpen his vocabulary he does crossword puzzles daily and consults maps with a kind of religious devotion, but his very favorite activity is poring over time schedules.

WHEN TO MUMBLE – *The Calgary Herald* – July 31, 1980

You haven't heard from Thaddeus McMurphyvisk for a while. That's because he has been compiling a dictionary of business laws and put-offs and other miscellaneous rules of thumb to guide the farmer. For instance, here is one Polish proverb every farmer should know by heart: *"Under capitalism, man exploits man; under socialism, the reverse is true."* Every farmer should know what a consultant is because the organizations and government departments that he deals with have hired many of them at $500 a day: *"A consultant is a status buck passer with a briefcase."*

McMurphyvisk has made a remarkable discovery about news sources. He defines them as:

Reliable source: A farmer you have just met.

Informed source: The farmer who told the farmer you just met.

Unimpeachable source: The farmer who really started the rumour.

Almost everyone knows Murphy's law: *"If anything can go wrong, it will"* But he has included in his dictionary McMurphyvisk's Commentary on Murphy's Law: *"Murphy was an optimist."* Similarly, there is McMurphyvisk's Corollary to the Law of Selective Gravity. The law: *"An object will fall so as to do the most damage." The corollary "The chance of the bread falling butter-side-down is directly proportional to the cost of the carpet."*

Delving into research stations, McMurphyvisk found three different kinds of activities going on there:

1. *Howse's Law*: Every man has a scheme that will not work.
2. *Gordon's First Law*: If a research project is not worth doing at all, it is not worth doing well.
3. *Etorre's Law*: The other line moves faster.

But all these activities can be ignored as there is The Golden Rule of Arts and Sciences: "Whoever has the gold makes the rules."

One of the farmers' most frustrating jobs is dealing with city businesses and not being able to decipher the reasons for their failure to act. Good old McMurphyvisk has translated some of these phrases that come back over the phone or in correspondence:

It's in the process: We forgot about it until now.

Take this up at our next meeting: That will give you time to forget.

Under consideration: Never heard about it until now.

Under active consideration: We're trying to locate all the correspondence.

We're making a survey: We need more time to think up an answer.

Let's get together on this: You're probably as mixed up as we are.

We can go over this at lunch: It's time we ate on your expense account.

Of course, mixed in with all this is Boren's First Law: *"When in doubt, mumble."*

McMurphyvisk has also polished up a couple of his own theorems he developed early on and included them:

McMurphyvisk's Theorem IV: Facts do not give us reality. On the contrary, they hide it. They present us with the problem. If there were no facts, there would be no problem, no enigma. There would be nothing to hide and nothing to discover.

McMurphyvisk's Theorem VI: Agriculture will be around supplying cheap food to other affluent segments of the economy long after they have eaten themselves into their graves.

Sometimes getting things done on the farm is just about as hard as it is in the average office, he discovered, and has noted down some of the doublespeak:

Activate: Make more carbons and add names to the memo.

Note and initial: Let's spread the responsibility.

Forwarded for your consideration: You hold the bag for a while.

Program: A project requiring more than one telephone call.

Project: A word that makes a minor job seem major.

Implement: Hire more people and expand the office.

The person to really watch for when he gets on a farm is the kind defined in *Segal's law*: "A man with one watch knows what time it is. A man with two watches is never sure."

WITH CRASH AND BANG, GRAIN HEADS OFF TO PORT — *The Calgary Herald* - January 6, 1981

A great many reports of a routine statistical nature hit my desk and are quickly despatched to the—ahem—wastebasket. This can be an unwise practice. Glimmers of stories creep out of the footnotes at times. The Western division of the Canadian Transport Commission of Saskatoon and the Alberta Gain Commission have both begun to forward figures on grain car unloading at the ports—reports, which I don't perceive as meaning very much. Doing some year-end file purging, I came across a handful of these reports and, idly leafing through them, received the impression that grain shipments are handled with reckless abandon.

It seems to me that when the people at the Canadian Wheat Board send off a trainload of grain they stand around wincing, with their eyes shut and fingers in their ears awaiting reports of smashes, bangs, delays, crashes, thumps and grinds all the way to the ports. They usually don't have to wait long. During October, November and December there was:

- Slam-bang: November 19th a Chinese vessel, Wu Ahi Shan, hit the dock and grain elevator at Prince Rupert and knocked off a loading spout, damaged two others and damaged the dock. One has not yet been made operational.
- Cr-r-r-unch: December 6th the Canadian National had a three-car derailment at Edson which tied up the line for 16 hours.
- Rumble, bang, screech: December 6th a CN derailment occurred in the Glenn Yard near Vancouver—and repairs took five days.
- Silence: The car dumper at Prince Rupert had to be taken out of service temporarily.
- More silence: Pacific No. 3 elevator at Vancouver was shut down October 6th through 23rd due to mechanical difficulties. It couldn't unload.
- Bump, grind: Pioneer terminal at Vancouver had an accident November18th which closed it down for four days.
- Clank, clink: November 15th CP Rail shut down main line for 36 hours for track maintenance.
- Swoosh: During the week of December 7th snow swept down the mountains and cut the number of car deliveries at the Alberta Wheat Pool terminal to eleven one day from the usual 140.
- C-r-r-rump, scre-e-e-ch: November 13th CN had a 21-car derailment on the Clearwater subdivision east of Red Pass Jct. which blocked the line nearly 48 hours.
- Thunder, caroomba: A derailment of seven cars on CP Mountain subdivi-

sion December 18 blocked the line 14 hours, adding to other problems.

- Groans, sparks: December 16th mechanical and electrical difficulties cut unloads at Pacific Terminals houses in Vancouver to 13 from 140.
- Poof, puff, grind: Dust and dumper problems at Saskatchewan Wheat Pool and Pioneer Grain elevators on December 18th affected unloading.
- Swoosh, slide, wrench: Floods in the Fraser Canyon December 25th knocked out CP and CN lines—and it was a week before any grain moved through again.

These are just a few samples of what happens after the seals are put on the cars and they are picked up by the way freight. There's no telling what might happen next to those grain trains.

Gifted with a rich imagination, Schmidt periodically gives it full rein to liven up his newspaper copy.

INTERNATIONAL DAIRY POLITICS COULD MAKE LIVELY FICTION — *The Calgary Herald* - January 13, 1983

One of the most complicated, subsidized, complex, secretive, political, cut-throat, distressing and crazy businesses in the world is the dairy business. I defy any person, even an investigative farm writer, to sum up the world dairy situation in less than 176,000 words. If I were writing a novel, I would have a ready-made cast of characters for a first-class piece of fiction. In fact, I can't see why W.0. Mitchell or Margaret Atwood don't try them out some time.

I suggest the title could be: *It's Always Better With Butter.* They'd have: Elsie the Cow: the principal female character. J.T. Graham, chairman of the New Zealand Dairy Board: the heavy. Eugene Whelan, the Canadian agriculture minister with the green cowboy hat: the victim. (They're trying to crucify Gilles Choquette," he rumbled recently). Francois Mitterand, president of France: the madman. John R. Block, U.S. secretary of agriculture: the hired gun.

Whelan is attempting to run interference for Choquette, chairman of the Canadian Dairy Commission (CDC) and his former executive assistant. Although Whelan says Choquette is secure in his job, cabinet ministers like Consumer Affairs Minister Andre Ouellet and Finance Minister Marc Lalaonde say Choquette has become a big embarrassment for the federal government. The word is that he has been given a chance to find a job elsewhere.

Things got interesting when Auditor-General Kenneth Dye began checking out the CDC. Choquette's curriculum vitae disappeared. Somebody checked an American university which graduated him. But the university could find no record of him ever being there. An investigation of him by the Public Service Staff Relations Board over the firing of Richard Tudor-Price brought forward a comment that Choquette had a degree from a Canadian university with a major in fertilizer.

Graham told a group of international farm writers in a speech at Rukuhia, that if a trade war in butter and other dairy products breaks out New Zealand will swamp the competition as "we are the best. We are the most efficient dairy farmers in the world — and we are survivors." And this is true. The average New Zealand dairy farmer milks 136 cows, thus making him the world's most efficient. He'd win any trade war with all that lush grass produced in his country.

In a sub-plot to the proposed novel, the New Zealand Dairy Board almost failed to survive an onslaught by N.Z. housewives. The board, in an attempt to stabilize the world market, bought 100,000 tonnes of U.S. butter last year and worked it through its own market with a proviso that the Americans withdraw from the world market. But it didn't tell its domestic customers what it was up to. Uncle Sam's butter wrapped in N.Z. wrappers was an old trick the Kiwis learned from the Canadians. When the N.Z. housewives discovered the pedigree of the butter—it had a different colour, taste and texture—they came out swinging rolling pins. This was despite the fact they were only paying $1 a pound whereas butter elsewhere in the world was selling for over $3 a pound. Graham says N.Z. isn't going to try that caper again.

The chairman was upset about "the recent dumping activities of the Canadians." When taxed about this grave international economic sin, Graham said the CDC had put a large tonnage of skim milk powder into Mexico. And so it had; it sold Conasupo 34,000 tonnes in a deal which the U.S. thought it had sewed up.

The order wouldn't have made much of a dent in the U.S. stockpile of several billion pounds, but the U.S. had expected the order as it had extended $1 billion in credit to Mexico to bail it out of its current financial difficulties. The U.S. also has a five-year stock of cheese on hand plus $2 billion worth of butter. Graham thinks the best way of handling that surplus is for Uncle Sam to give it away to Americans. Delivery of the Canadian powder was completed to Mexico before Christmas via 460 railway cars and by ship. Ironically, the cars travelled through the U.S.

The CDC also sold 11,000 tonnes each of skim milk powder to Algeria and Iran. The sales cleaned out CDC storage stocks—but at a price. The CDC sells to the domestic market for $2,600 a tonne but the foreign market price is only about $900 a tonne. The actual sale price is a CDC secret. The federal government was picking up some of the loss, but recently shifted the whole load to CDC—the load this year being about $127 million. But that isn't the whole burden of CDC troubles. The Mexican government is technically bankrupt and dairy leaders are sitting on tenterhooks waiting to see some pesos appear.

The Kiwis thought much of the world butter surplus would disappear if the European Economic Community (EEC) sold its surplus of 300,000 tonnes to Russia. But the EEC Council refused to sell because that would have meant butter would be selling Moscow stores cheaper than in West European stores—and the council was likely to be beset by European housewives with rolling pins. With the deal cancelled, France vetoed the sale of 87,000 tonnes of N.Z. butter to the British market. While N.Z.'s Graham may cast slighting remarks at Canadians for dumping surplus product, it must be remembered Canada has a wonderful milk supply management system in force. It is so wonderful it is going to cut production 15% and require dairy farmers to pay $50 million more in over-production penalties this year.

While Whelan has boasted about the great job supply management has accomplished, it hasn't really worked because dairy farmers made money exceeding quotas in the last couple of years and nobody tried to stop them. They did this in the face of declining butter consumption. Now that the day of reckoning has come, they don't find themselves in a much better position than those in N.Z. where no quotas exist. Its dairy farmers are producing wide-open, going as hard as they can.

And, that ladies and gentlemen, is the stuff fiction is made of. It would test a writer's credibility to pass some of this stuff off as reality. In the view of some these revelations, the McMurphyvisk plan for world agricultural survival begins to sound better all the time.

Agriculture ultimately permeates all human activities everywhere, especially in Canada where Schmidt draws attention to the link between pucks and bacon.

GAINER'S GAINS A GRETZKY
– *The Calgary Herald* – May 9, 1984

For the open session of its annual meeting in Edmonton, the Alberta Pork Producers Marketing Board invited the presidents of the three pork-packing companies left in Alberta to a panel discussion on the state of industry. Leo Bolanes, the new boy on the block, just in from Midwest Packers Ltd in Tupelo, Mississippi, appeared to be in an enviable position. He has a contract with Peter Pocklington to run the Gainer's Ltd. plant in Edmonton. Peter Pocklington? Isn't he also the man who has Wayne Gretzky under the $20-million contract to play hockey for him on the Edmonton Oilers hockey team in the NHL? Yes, indeed.

The packer in the most unenviable position is Ron Faithfull, president of Burns Meats Ltd. He was talking gloom and doom and instability. Then there was Garry MacMillan, the former assistant manager of the pork board, who is now president of Fletcher's Ltd., and who now sees a major opportunity to expand plant facilities to make Alberta a hog butcher for the Pacific Northwest of the United States.

As the speakers droned on and on, I kept drifting off into the euphoria of the Stanley Cup playoffs. I envisioned Pocklington's man, the Great Gretzky, being madly idolized by the crowds and sports writers as he helped the Oilers to skim from one victory to another. When the face of Bolanes came into the hog producers' arena it appeared to have on a big hockey helmet as he stick-handled his way through the Faithfull and MacMillan lines. When the euphoria of the game took greater hold, I could see Ed Schultz, manager of the pork board, skate by in a black-and-white striped referee's sweater being booed by the fans for missing an offside by MacMillan. The voice of Howlin' Howie Meeker of New Hamburger changed to an "oink, oink."

"They've got Bolanes in a corner—and he can't get at those hogs," squealed Meeker jumping up and down. There was a whistle on the play by linesmen LeRoy Fjordbotten, the Alberta agriculture minister, and Harvey Buckley, chairman of the Alberta Agricultural Products Marketing Council. They called a face-off at centre ice. Once more Howlin' Howie came through the image: "At the face-off Bolanes made a breakaway for the Gainer's hog team. He took a rush at the press, which quoted him as being starved for hogs by the pork board and he was going to reduce plant staff.

"Now the pork board put 37,000 hogs on the market last week and slid 14,000 into Gainer's net—the biggest, yea, the biggest, kill in the history of the plant," yelped Meeker. Back in the dressing room, Bolanes was denying he had come to Gainer's as a quick turn-around specialist. "Pocklington offered me a three-year no-cut contract to come here and run Gainer's. Now he has changed the game plan. I have a 10-year contract which calls for 10% of the gate plus salary. I'm on the fast track for more profits five years down the road."

Back at the pork board arena, there was a great cheer as Bolanes skated out at the beginning of the third period. "I could kill 5,000 hogs every day of the week. All I want you to do is give them to me." He picked up a penalty on that one. "No, I mean sell them to me," he said amid roars of laughter from the crowd. But he wheeled and took a slap shot at Burns and Fletcher's. "I need 5,000 hogs a day and if I could get them I'd be the most efficient packer in the West. Maybe we should set the price the week before to ensure delivery to me!"

Hey, give us an instant replay on that. Did I see it? Yes, I did. There was the ghost of the late Charles McInnis whooping it up along the boards. In the 1950s when he was president of the Ontario Hog Producers Association, Charlie went into a meat packers' meeting and made this same suggestion: that the price be set in advance. The packers threw a five-to-one power play at him for this radical suggestion—and he retreated to organize the Ontario Hog Producers Marketing Board.

As the game was nearing the end the cameras focussed on a picket sign: "I Need 80 Cents For Hogs. You Pay Me Only 66."

This was when a slashing match broke out at centre ice between Burns, Gainer's and Fletcher's. When Pocklington inked a contract with Gretzy for $20 million—and other team owners did the same with their super-stars—they turned around and quadrupled ticket prices. And there wasn't a squawk from a fan. They boosted the television rights 10 times—and still no squawks. But instead of boosting pork prices the next day to give producers 80 cents, the three presidents started talking about competition, telling the producers to look down the road 10 years, not deliver all the hogs on Wednesday and flood the killing floors and pointing fingers at Quebec.

It was a losing game from there on!

In addition to providing his in depth analysis of human activities and creations through an agricultural perspective, Canadians can thank Schmidt for saving them more than once from bunglecrats

FLUFFY IDEAS – *The Calgary Herald* – October 23, 1985

About once a decade some fluffy-headed civil servant comes up with a brainwave that gets him almost laughed out of Ottawa.

Back in the seventies somebody thought it would be a first-class idea to place jailbirds with old folks. The senior citizens would be able to talk kindly to the cons and change them into thoughtful, caring citizens. For this the aged would be paid a supplement to their old-age pensions. The whole idea was aborted when almost every writer and Conservative in Canada laughed uproariously in print. Being a practical joker of sorts, I was disappointed when the idea dried up because I wanted to arrange for my sainted mother, then 78, to have a chain gang of five or six prisoners to look after her garden in summer and shovel snow in winter. They would be supervised with a blacksnake whip by Thaddeus McMurphyvisk.

The new, wildly hilarious idea for the 80s to come out of Ottawa is a product of the fertile mind of Duncan Ellison—and he's in charge of the Transportation of Dangerous Goods Act for the Ministry of Transport. He is the man, who as of July 1, 1985 has required everyone mailing a parcel to sign an affidavit signifying that such parcel does not contain any dangerous substance or material, or non-mailable matter as defined under the act or the Canada Post Corporation Act. Each person swears he fully understands the extent of this declaration, which is based on a book almost four centimetres thick.

Not long ago I received a parcel from a farmer up north with one of the yellow affidavits on it. I called the post office customer services division representative and asked what it intended to do with the parcel as I suspected it contained "dangerous material." This stopped the customer services worthy in his tracks. No regulations had been written on what to do if a client receives a package and suspects it contains dangerous material.

"Have you opened it?!!, he demanded.

"No," I replied.

"Then how do you know the material is dangerous?"

"I have received material from this address before," I replied, "and it is very dangerous. The correspondent advocates changing the world; maybe even blowing it up. He wants new farm policies and leads a movement to start revolutions and new monetary policies. Very dangerous right-wing stuff."

"Call the police", he volunteered.

"No," I demurred. "This isn't police business. You forced your client to sign this affidavit. Now you want to shuffle off all responsibility in the matter. Wait until Christmas rush and you'll have riots on your hands when the customers have to wait in lines half a mile long to mail gifts."

"Well, call post office security and see what they have to say," he blustered at my loud guffaws occasioned by his discomfiture at having been caught out. I called post office security and the worthy there was likewise impotent and naked at being unable to inform me what the post office intended to do with my parcel.

"Are you going to open all parcels to check out the affidavits to ensure the public no dangerous material or prohibited substances are in them?" I pursued.

"No, we have neither the time nor staff," was his reply.

There are 1.5 million people out of work. It would assist government employment statistics to hire people to open the parcels and it would be more beneficial than requiring all the clients to sign affidavits saying they had read and understood a 522-page act. He grew uncomfortable and hostile at my guffaws and he too passed the buck. "The MOT foisted this on us. The man to put the finger on is in Ottawa. Call him." Which I did forthwith.

On the line came the redoubtable Duncan Ellison, the man who put the bureaucratic nightmare in motion at all post offices in Canada but had no idea what to do at the receiving end if an addressee suspected a parcel was full of dangerous material. When I started to guffaw uncontrollably at his suggestion that this would prevent persons from mailing such material, he bristled: "This is a serious matter."

"Yes, it is for you," I said. "Every parcel from now on with dangerous material in it will have your signature on the affidavit; or your boss, Transport Minister Don Mazankowski."

There was more sputtering on the other end: "You could go to jail for this."

Now that gives me an idea. We're practical jokers in our family. Next time I send some perfume in the mail I'll sign my son, John's name and address on the package. I've been trying to hang a federal rap on him for a long time, and he on me!

The MOT has opened the way. My sides are hurting from laughing. Almost as much as if my dear old mother could have laughed if that chain gang of prisoners working for her was being lashed by good old McMurphyvisk.

Schmidt cringes in horror at the prospect of being told, "he's in a meeting" when he calls a farmer to check out a story. He has been laughing like a hyena ever since it became fashionable to equate humor with hate laws. And he enjoyed seeing practical jokers serving a notice on Prime Minister Trudeau to come and cut the stinkweeds on "his" property at Patricia, Alberta.

THREE-PIECE SUITS MAKE INROADS INTO RANCHING — *The Pincher Creek Echo* - February 21, 1989

An amusing yarn about American business methods made its appearance recently. In an attempt by the socialist government of Algeria to show how decadent the capitalistic U.S. cattle and oil business is, the state-run television began showing re-runs of "Dallas." The objective was to discredit the American system. However, in a nation of 17% unemployment and anti-government riots in the streets, the TV series on the freebooting Ewing family of South Fork Ranch backfired. The Algerian people were fascinated by the clear and simple message of "Dallas": Get rich. They love the Ewings and pay no attention to the critics: the press and clergy.

Apparently the government finally got the message: the Algerians like the way South Fork and the Ewing Oil Company are run as they make people rich. The government hasn't delivered on its promise of a happy and prosperous future for all segments of the population. The people got sick of the paternalistic and authoritarian state. "Dallas" is given credit for causing the riots.

But, inasmuch as "Dallas" style is appealing to outsiders, some fundamental economic changes have begun to show up on some of the big ranching spreads in the U.S. What happens in the U.S. gradually gravitates to Canada. In fact, the trend has probably started. The biggest change has come at the best-known of all U.S. spreads, the King Ranch of Texas. For the first time since its founding in 1853, three "outsiders" have been brought onto the board of directors to bring the corporate realities up-to-date. In the early 1980s the ranch reached 1.2 million acres in size—with holdings in Florida, Kentucky, Pennsylvania, Australia, Brazil, Argentina, Spain, and Venezuela. But it has now been scaled back to less than a million acres, with the sale of holdings in the last four countries. The new board cut the work force in half to 325. Some of those let go had been fourth-generation employees.

The ranch sold $1.5 million in horses and cattle (it originated the Santa Gertrudis breed) at its 36th annual sale but it continues to be pinched by lean times in oil, agriculture and real estate which have hit other U.S. corporations. The new directors were brought in from the corporate boardrooms of Maxus Energy Corp., Kimberly-Clark Corp and LaSalle Partners. In future, a call to public relations director of the King Ranch could bring the reply: "He's in a meeting."

"Oh, gawd!"

And the ranch has gone into fish farming. Whether the infusion of non-agricultural outside blood is the answer to cattle ranching remains to be seen.

Many ranchers will believe it is on par with our Texas cousins electing a Dallas corporation lawyer specializing in taxes, as president of the Texas Cattle Feeders Association. He is Clark Willingham, 42. Starting out as a big-city-type lawyer, he married into the Hitch feedlot operation at Guymon, Oklahoma, one of the biggest in the U.S. (400,000-head turnover annually). He was sent by Hitch and other clients to Washington as a lobbyist to soften the effects of the 1986 U.S. tax bill, which removed the tax shelters for thousands of investors in the feedlot business. Willingham then became interested enough in the feedlot to have a go at serving as a director in the feeders association. His big priority will be to convince feeders that if they do not continue to support an advertising and promotion check-off of $1 a head, beef will lose out to other meats. The check-off is coming up for a review and vote in May. His other corporate thrust lies in the production of branded beef—a name such as HitchingPost Beef.

The other corporate reality of U.S. farmers and ranchers is their reliance on the scientific world for improving their product, and increasing its saleability in a world that beef is being too often by-passed by "health-conscious" consumers. Their sons are being sent to university to specialize in animal science, nutrition and marketing. People will pay attention to Dr. Dean Miller, a graduate of Texas A and M University and now a marketing specialist with the University of Georgia, when he says: "Beef is a most nutrient-dense food. It would take 50 pounds of cabbage to equal a three-ounce serving of beef." He foresees the day when retailers will include nutrient labelling on branded beef the same as on other trademarked foods for the benefit of finicky consumers. He agrees with Willingham that branded beef is nearly a reality. And, he concludes: "One supermarket chain has begun to market branded beef in vacuum packages—and the meat will stay fresh in the refrigerator for two weeks."

Over time, as Schmidt can personally confirm, there are definitely changes in what society (and newspaper publishers) will accept as humorous.

LAUGHTER HAS NOW BEEN POLITICIZED
- *The Olds Gazette* - June 28, 1989

Thaddeus McMurphyvisk and I have just completed a monumental study on what Canadians are permitted to laugh at these days. Our conclusion: Not very much. Things that we can tell jokes about now seem to be graded on a scale that ranges between "innocent," "twitting," "tweaking" and "hate literature." Anything that causes more than a smile is likely to be censured. The ranking appears to be done by a shadowy modern-day star chamber of hard-faced humourless thought police heavy weighted with ethnics and UNREAL women.

They are attempting to get McMurphyvisk to change his name because they allege it has racist overtones and holds Scotchmen up to ridicule. The same kind of jesters who have set the unofficial laugh standards—and, what is worse, have gotten away with it—are the same people who have forced the minister of national revenue to prohibit the entry of 500 books into Canada. One of the banned books is Joe Miller's Joke Book, which was the book comedians used to get their laughs. Half the jokes began with: "There was this Jewish rabbi and this Negro preacher walking down the street." Then followed a knee-slapping joke, which would be repeated the next day in the local poolroom and that afternoon at the local Farmers Union meeting.

I sorely miss the jokes at the farm meetings nowadays; everyone is terrified to tell any because the thought police may have a hidden microphone. Alas, this type of joke is not acceptable any more. It is classed as hate literature. About the only types of humour which can be retailed these days without pricking thin Canadian skins are lampoons of Tory politicians, faith healers, bathing beauty contests and stuffed shirts who are honoured for achievements in their pedestrian careers. Of course, it's well nigh impossible to try to suppress laughter among human beings—or hyenas. This has meant certain types of humour have gone underground until they become socially acceptable again. What has gone underground is black humour of the kind that could get a Soviet citizen sent to Siberia pretty quickly before perostroika.

Sorting out the humour from the bad taste is the reason many daily papers have appointed ombudsmen. The ombudsmen are nice guys—but they don't do anything. They front for editorial staff, who have neglected to interpret the parameters and draw howls from readers. Ombudsman Jack Briglia of the London Free Press got yanked into the barnyard recently when a hog producer complained the wiseacres on the paper's staff didn't treat him with respect. The occasion John Gough, 34, of Mount Brydges, complained about was when his family was honored by the London Chamber of Commerce with the Farm Family of the Year award. The award was designed to foster better understanding between rural and urban communities in the London area. Gough figured it was swinish for the Free Press newsroom wags to characterize him as the stereotypic hick. He was irked with a carelessly written story by Howard Burns, a greenhorn farm writer, which misstated his qualifications and address and referred to his two sons but not his wife and business partner, Enid. He figured this omission was a slap in the face to her.

His curriculum vitae revealed he had an agriculture degree from the University of Guelph and that he was a member of a barbershop chorus. He was more than upset when Burns used these facts to amuse his readers by calling him "a farmer with a passion for barbershop singing, who hit a high note with the London Chamber of Commerce." He intends to use that voice to "better inform city slickers of the contribution farmers make." The headline writer referred to him as a "singing hog raiser." But the cutline under a photo incensed him as it read: "When John Gough isn't slopping hogs, he sings in a 50-member barbershop quartet. Farming, however, is his first love." Gough told the ombudsman that the paper's use of such a term as "slopping hogs" and "city slicker" are insensitive and out-of-date and torpedoed the objectives of rural-urban understanding.

The ombudsman agreed the staff came out smelling like that stuff Gough spreads around the fields. But he forgot one important part of the story: that Gordon Wainman, long-time farm writer at the Free Press had been given the bum's rush when he objected to the editorial department's similar frivolization and trivialization of many of his serious stories on his beat.

Undaunted by any prohibitions of mirth, Schmidt finds that even landmark constitutional debates have humorous elements.

HERE'S WHAT WE HAVE LEARNED ABOUT CONSTITUTIONAL DEBATES
— *Western Stock Growers Newsletter* – December 1992

By the time you read this, that referendum and the civic elections will be over. There were a great number of praying politicians going after the vote in this go-around because they realized the initiative and recall legislation might be just around the corner. To avoid this, many imported a politician's prayer from Seven Persons, Alberta. It is: "Teach me, O God, to utter words that are tender and gentle, for tomorrow I may have to eat them."

That's almost as good as the one-liner Arthur (Bugs) Baer got off in one of his columns which ran in Hearst papers in the United States over 50 years ago: "Europe is a place where they name a street after you one day and chase you down it the next." He was the humorist who also wrote: "We must make Europe pay for the last war to prevent it from affording the next one." Baer would have given short shrift to all the millions of words wasted on the Canadian constitution when the politicians should have been drinking beer—had he been alive now: "Of all our constitutions I like, the first one is best."

And will we get the truth out of more senators? I heard a description of one senator: "The senator was as frank as a postcard in red ink." A new senate could be refreshing. The one now reminds me of an apocryphal story of Cactus Jack Garner of Texas, who said (and I paraphrase him): "Two brothers were born to a family in Alberta. When they grew up, one ran off to sea and the other was appointed to the senate. Neither was ever heard from again."

Old Dr. C.O.N. Buller once said that if he stopped to read any of those bills rubber-stamped by the Canadian Senate, "I would jack up my rear wheels and count my money."

And Dr. B.U.L. Conner wondered why more MPs don't go as cuckoo as 12 o'clock in a Swiss clock factory.

Abe DeFewgilty is quoted as saying: "When anything is wrong in Ottawa, they start a new department for it."

Aunt Sadie Glutz thinks: "We're in an awful fix unless prosperity takes a turn for the better."

This blockbuster came from J.P. O'Flasky: "There is nothing in the bill of rights that entitles a politician to a night latch key to the federal treasury."

When they all came home to sell the constitutional package from Charlottetown, they saw tell-tale signs that the economy was in a tailspin:

- When they went out to dinner, they saw more people eating off each other's plates.
- When driving along No.1 Highway they saw former employees of PetroCanada hitch-hiking back to Montreal.
- At the local supermarket more and more mothers were telling the kids to put things back on the shelves.
- Teen-agers who wouldn't have been caught dead doing it before are now collecting beer bottles.
- More people are hanging onto their jobs—even though they hate them.

The whole constitution looks to me like the time they were going to hang a fellow at Medicine Hat. As the fatal day approached, the sheriff asked the prisoner if he had one last wish. The condemned man replied he'd like to inspect the gallows. The sheriff allowed that that was all right. They slowly walked around the gallows and the man inspected the structure quite closely and when finished turned to the sheriff and said: "Ya know, sheriff, I'll be honest with you. The damned thing just don't look safe."

A lot of the politicians reminded Bugs Baer of his second day in kindergarten: "I once brought this teacher a big red apple. She took it. I haven't trusted women since."

"What I'd like to have seen the politicians do was postpone the whole debate for five years—just like a bath in cold weather."

You may have heard stories and one-liners like this before. Unless you want to follow the old political tradition and steal them outright, they may cost you $12.95, as they are all found in a book recycled by two Texas Democrats, Chuck Herring and Walter Richter. The book is called *Don't Throw Feathers At A Chicken.* However, I have a special deal for anyone getting a speech together: you can steal anything you want from this column. And for good luck here's another one thrown in:

A politico was rushed to hospital in Cochrane after being bitten by a rabid dog. When the doc came in he was furiously scribbling on a legal pad. The wound was not life-threatening so the doc told him he could stop writing his will. "Oh," replied the politico, "I'm not writing my will. I'm just making a list of all the people I want to bite."

(All these stories are certified politically correct by Dr. Raffath Sayeed, acting head of the Alberta Human Rights Commission.)

Schmidt uncovers definitive proof that: "One man's meat is another's poison - one man's humor is another's headache."

GIFTS TO KLEIN AND TRUDEAU GOT THEIR GOATS — *The Fort Macleod Gazette* - June 8, 1994

Premier Ralph's refusal to accept a pygmy goat he won in a raffle at Brownfield is reminiscent of Pierre Trudeau's refusal during the Nixon years to accept a building lot at Patricia to establish a Western Meech Lake. Baa!

In March students at the Brownfield school held a "Give Your Neighbour A Headache" raffle to finance a new public address system. The winner (miraculously due to the "luck" of the draw): Premier Ralph Klein. The prize was a loveable little billy goat with mutton-chop whiskers like Rod Love his executive assistant. There was symbolism. Premier Ralph had threatened rural school closures. Love told the kids his boss, Premier Ralph, didn't have time to come and pick up the prize—not even to install it in the opposition benches. The kids realized they had succeeded in getting his goat and giving him a headache. Baa!

This same lack of appreciation of a joke was present in the great Pierre Elliott Trudeau—and I doubt whether it rated a reference of his refusal to accept a bucolic gift from Alberta in his current best-selling autobiography. It all started when Albert Ketchmark of Bow City and Jack Horner's official agent, Eugene Kush of Hanna, faced a situation which was giving them a headache. Kush, a laudatory rural lawyer, was called in to help settle the estate of Ketchmark's mother. The estate included a building lot which she had bought on speculation during an oil boom in the 1940s. Located in beautiful downtown Patricia, it had proved to be worth less than the $125 it would cost to transfer the lot.

The pair decided to hold a "Give Your Neighbor a Headache" raffle, with the lot as a prize.

At that time President Richard Nixon of the U.S. was talking about setting up a Western White House in California to cut his commuting time. Therefore, the Ketchmark lot was won by Trudeau and he was sent a note suggesting he set up a "Western" Meech Lake on the bald-headed prairie at Patricia. Their best intentions and good will were unilaterally overturned by Ivan Head, a Trudeau executive assistant, who returned the deed and pointed out the prime minister couldn't accept such thoughtful gifts. The only trouble was Kush couldn't figure out a procedure for taking back the gift—and nothing happened until a member of the legislature press gallery in Edmonton saw Trudeau's name on a list of Alberta properties to be sold at a tax sale some years later.

The reporter had a scoop. Trudeau had a fit. Jim Nesbitt of the Brooks Bulletin had a brainwave and paid the $17 in arrears.

Tom Musgrove, reeve of the County of Newell, responded with a notice to Trudeau to come and cut the weeds on his property. The upshot of accusing the prime minister of the heinous crime of not cutting the stinkweed was that lawyer Trudeau despatched a couple of Horsemen (RCMP) to ride out from Ottawa and tell lawyer Kush to get rid of this headache—or else...he might bring him before the bar association.

Sometime later a want ad appeared in the paper: "For Sale—A building lot in Patricia which once belonged to Prime Minister Trudeau. Important historical site. Reasonable." The lot was sold to the Roman Catholic Church—at which point it became tax-free. At least, that is the way I heard this vignette of history in a bar last week.

Ever mindful of the actions taken by leaders of various groups, including governments, Schmidt's mission is to encourage his readers to carefully assess how these actions impact the individual.

IN THIS CRAZY NEW WORLD, TURN OTHER CHEEK — *The Settler Independent* - December 21, 1994

Everything seems to be going backwards in this crazy world. More than 50% of grain growers' incomes comes in the form of cheques from government rather than cheques from grain sales. Grain farmers have to beg unionized dockworkers to work overtime to move their grain to world markets to feed the people. Whereas many growers used to genuflect at the mention "Canadian Wheat Board," young aggressive growers across the West are now giving the CWB regular pummellings.

The post office can't—or won't—deliver weekly newspapers to subscribers in the week they are published any more yet it is horning in on every other business in sight. It is setting up junk mail retailing centres in the big cities. Next thing we know, it will be in the used car business.

The Hamlet of Chancellor has undertaken an initiative to become "Canada's first politician-free zone." No, that initiative hasn't been published in the Alberta or Canada Gazette but a small ad has been taken out in the Village of Standard Community Calendar notifying the public of that important step.

In a speech she gave in Calgary in January to the Alberta Institute of Agrologists, the former Alberta consumer and corporate affairs minister said:

"There are no Canadians any more. We are all members of special interest groups." The special interest groups spend a lot of their time picking on each other. Numbered among the special interest groups are governments and seniors. The governments have been especially active in plucking the seniors in many ways they never imagined they would be plucked.

I find this to be ironic in view of the fact the Hon. Members of the Legislature in Edmonton hold a prayer session every day before the plucking starts. However, they have apparently overlooked the Fifth Commandment which says: "Honour thy father and thy mother that thy days may be long in the land that the Lord, thy God, giveth thee." There is another adjuration in the Good Book somewhere that if somebody gives you a belt in the chops the best policy is to turn the other cheek.

After the most recent round of clawbacks our household decided to heed the Biblical adjuration and help bail out needy government with a week's wages. To prove the point that some of us started out in a humble way, the week my wife and I chose was the first week we entered the Canadian work force. Her cheque was for $7 and mine was for $2.50, believe it or not.

The cheques were tendered to the government July 15 through the MLA for Drumheller, whose name escapes me at the moment. However, it wasn't until Nov. 3 they reached the hands of Michael D. Faulkner, senior financial officer of the Alberta Treasury. We hadn't reckoned on an apparent government policy that it is more blessed to give than to receive. It's easier to write cheques than receipts. Receiving money from those of good will seems to bewilder the powers-that-be in this world that's going backwards. The MLA sent the money back without any explanation. What to do?

It isn't often I get a chance to meet Premier Ralph but when I got lucky a couple of months later in Edmonton, I observed to him his government appeared to have cut the deficit so much it was no longer short of money. As proof, I submitted the story about the two returned cheques in my pocket.

He was not daunted by the action of the bureaucrat Faulkner, but laughed and said: "Christ, I'll take your money if nobody else will. Just send it to me in the mail." And, like the Little Red Hen, he did. He took our money. And so, with the season of good will, cheer and giving upon us, wouldn't it be appropriate for each citizen of Alberta to play Sanity Clause in this crazy world?

Send the government treasurer a Merry Christmas. And give thanks to this country which has been so good to us all.

Schmidt discovered early in his career that his readership covers a wide spectrum. It ranges from those who profess a distain for manure columns to those who delight in homespun tales of rural life. There is an academic constituency who groan at some of his outrageous musings and corny comments, as well as a loud, raucous bunch of cowboys who appreciate sly innuendos. There are also those who have a hard time contending with his muck-raking and occasionally there are some who send him plain brown envelopes to help spill the beans on the shysters.

As he progressed through life (one in which he was driven crazy not knowing what dilemma would face him each morning) he began to regard the world in a whimsical manner. In his writing, he uses humor and satire (not always successfully) to drive home complicated concepts in agriculture. His stock in trade was to write bizarre or offensive lead-in paragraphs to capture the readers' attention and then to carry on reading the whole column.

Schmidt's first mentor in using humor and satyr was his uncle Mooney, who convinced him that they were good tools for keeping one's writing from being dull and pedestrian. The use of these linguistic tools had the joint affect of either having readers leaping with glee or turning purple with rage.

He maintains that his is a life of fun, games and high dudgeon. He further claims to be astonished that publishers actually pay him for it. In reality, however, Schmidt never stops thinking about what next to write. He spends almost every waking moment engaged in collecting information to share with his readers. His relentless searching sends him around the barnyards and boardrooms of the world as well as keeping a sharp eye on what other writers are writing.

CHAPTER SIX

Food For Thinking

THE EVOLUTION OF AGRICULTURAL WRITING

Over the past sixty years, the tremendous innovations in agricultural technology changed farming from back-breaking labour into sophisticated business operations. Like farming, farm writing too evolved into very different opportunities for writers in the agricultural field than those available in 1938, when John T. Schmidt first started writing columns for his family's newspaper.

From the beginning of the twentieth century, Canadian farmers benefited greatly from the information provided by the phenomenal number of papers and magazines that were dedicated to agriculture. However, by the early 1970s current events became more divorced from things agricultural as the farm population dropped in half. In the decade between 1980-1990 the schism widened as economic changes moved still more farmers off the land. By the turn of this century, all the big urban newspapers had eliminated farm columns from their pages, reflecting that fact that farmers now make up less than 3% of the national population.

Periodically, Schmidt provides his readers with an "insider's" view of farm writing, the role that farm writers play within the context of the larger community, a glimpse into their professional organization, and his assessment of some of the other writers in his field. Possessing an unwavering commitment to Canadian agriculture and his profession, Schmidt provides a pithy summary of the developments he witnessed in both over time. The following collection of columns is a sampling of his views on the various aspects of the media as they relate to agriculture.

A SNAP SHOT OF FARM WRITING

The history of farm writing in Canada followed the pattern of settlement. The nation was supportive of agriculture as agriculture was the only game in town from which the business community could create new money. Therefore, the pioneer news weekly and daily papers as well as news magazines gave plenty of space toward assisting agriculture to become Canada's largest industry. All farm writers were essentially newsmen. As the population grew, the daily papers became more interested in reporting the "newsworthy" polyglot happenings of urbanites. Those who managed the news presentations began to regard farm news as "pedestrian." It was harder to find space on the front pages for the pedestrian unless there was a catastrophe. But the papers couldn't avoid carrying farm news between catastrophes so they developed specialist farm writers to maintain readership.

Generally speaking, the farmers were greatly appreciative of the work that farm writers did in comforting the oppressed and oppressing the comfortable, not to mention exposing the guilty, provoking the greedy and mocking the powerful. Nonetheless, newspapers continued to squeeze out even pedestrian news content. The post office discontinued same-day mail service for daily papers. Nietzsche had declared God dead and the daily papers took this one step further by declaring agriculture a dead horse in the 1970s. Fortunately, farm writers didn't disappear. They remain useful to agriculture by writing for a large number of specialist publications, government, radio, television, and, in more recent years, as employees of agribusiness.

To forward their professional development, farm writers have an association in each province, all of which are under the umbrella of the Canadian Farm Writers Federation, except for Quebec. The English and French farm writers have not yet achieved confederation as each claims to be able to drink the other under the table. Unable to reach a consensus, the Quebec farm writers have chosen to go their own way. The CFWF was set up after the Second World War. As there is a wide spectrum of political thought among the members across Canada, the federation is maintained by a fraternal agreement whereby members never argue about their constitution when they should be drinking beer. The membership of CFWF has experienced a demographic change in that more than a third of them are now Spritzer-drinking women.

INFORMATION CRUCIAL TO FARMERS
-The Kitchener-Waterloo Record - 1956

In a recent column we commented on the fact that some farm leaders now believe a lack of communications between the provincial offices and the man on the back concessions is responsible for some of the opposition to marketing schemes. We speculated upon the possibility that the farmer might not be taking advantage of all the information available to him. Phil Novikoff, public relations manager of Canadian Industries, Ltd., has a different slant on the matter. He told this story to a recent meeting of the Canadian Farm Writers Federation:

He was reminded first of information that used to reach his farm home in Saskatchewan. "In the morning we lit the fire with the *Family Herald and Weekly Star*," he said facetiously. "Then we relaxed for a spell with Eaton's catalogue. In the evening we huddled over a two-tube battery radio set with the earphones glued to our ears. That was about the extent of the news from mass communication media that was extended to the farmer in giving him information to help him run the farm. Some families, of course, preferred the *Free Press Prairie Farmer* because the farm wife received a larger bonus from the subscription salesman."

Mr. Novikoff then went on to outline the amazing change in news service available to farmers and others. "The oil and propane stoves have edged out the *Family Herald and Weekly Star* as kindling material—so much so that this veteran publication went 'slick paper,' shortened its name and now about the only thing it's good for is reading. Radios are so commonplace they are even found in milking parlors where contented cows chew contented cuds while milking machines extract increased gallonage as the speaker blares forth rock and roll. And now TV has come onto the rural scene."

In Canada today there are 48 English and 9 French farm journals being published with a total circulation of 3,000,000 and a possible readership of 10,000,000. There are 84 English daily papers and 12 French with a combined circulation of 4,000,000. Because of expanded circulation in rural areas many of these have added farm writers to their staffs. At a large proportion of the 187 radio stations and 39 TV stations across the country only 4 have daily farm broadcasts. The CBC is making a splendid effort "to make millions of listening and viewing farmers into wealthy men," he said.

"The fact is communications have reached a somewhat phenomenal stage in Canada. No businessman, no industrialist, no tradesman, no professional person gets as much information on how to succeed beamed at him as the farmer. A manufacturer must hire experts or pay stiff fees to consultants to find out how to turn out and sell a product so he can make a profit. But the farmer has the best specialists in the country at his command—without having to pay a nickel for their valuable advice," said Mr. Novikoff.

In addition to this the federal and provincial Departments of Agriculture and dozens of commercial firms annually publish mountains of literature on better farming methods. Hundreds of speakers are touring the countryside addressing meetings on how to increase profits. Free movies are available. Field days, plowing matches and country fairs are for the farmers' benefit. And to top all this there is the local agricultural representative, that walking encyclopaedia of farm wisdom.

When all these sources of information are added up, Mr. Novikoff wonders: "Is the farmer inundated and confused by this welter of informational benevolence? Does the farmer have time to read, hear and see all the information available to him?" Somehow, we cannot but think that there is a lack somewhere. We cannot but wonder, as the farm leaders have wondered, how the farmer can sort the wheat from the chaff. One thing which still influences the farmer is the "talk" he hears around the feed mill, the egg grading station, the pool room and the general store. It takes a smart man to sort it out, yet he cannot live unto himself these days.

FARM AND RANCH REPORT
– *The Calgary Herald* – March 7, 1967

The Calgary Herald's Farm and Ranch Report will go into more than 130,000 southern Alberta homes today. In addition to the 92,000 circulation of *The Herald,* it will be distributed by mail to 40,000 farmers and ranchers and other people who are at the helm of Alberta's $794,000,000 agricultural industry. The circulation department says this will be the largest coverage special feature supplement in the paper's history.

I was pleased to be able play a key role in the preparation of this report. I approached it from the point of view that it was part of my centennial effort. Perhaps researchers for the next Centennial celebration might dig it up and laugh and chortle about the way agriculture was carried on in 1967. Or perhaps they will find agriculture in 2067 can benefit from methods practiced today. Who knows? At any rate it was a lot of fun digging up some 85,000 words to fill in the spaces between the ads. That is the number of words in a small novel. It's not likely many people will read all of the words because they won't be interested in all the topics covered. I attempted to range as far a field as possible and, even so, left untouched many fields in Canada's largest industry.

The agriculture beat is one of the most exciting on any paper these days, in my opinion, because of the giant strides agriculture has made in the business of producing food. There is nothing static about it. There is something new and different and amazing to learn every day. It is my hope that each and every one will find something of interest in both the editorial content and the advertising columns.

Incidentally, every time I see an advertising salesman I take off my hat. He or she is the means of providing the revenue on which a newspaper thrives. In my weekly newspaper days selling advertising was one of the multiplicity of jobs upon which I was called to do—and I hated every moment of it. It requires a person of a special temperament to sell ads.

Speaking of ads, there is one on page 20 which should jar the parochialism of every good Calgarian and true. Inserted by the Edmonton Industrial Development Department and *The Edmonton Journal*, it definitely dashes the long-held belief of most Calgarians that Edmonton is merely a big mirage in the north end of the province! Page 20 also carries a story about a man who lived in a very much less affluent society than Canada for 18 months. A.W. Beattie, the Calgary district agriculturist, returned last year from a FAO assignment in two African countries. He describes how the United Nations groups plan to come to grips with the problem of underfed people. His story can't be told too often.

We Canadians have been doing a great deal of talking about high food prices in the last year. The figures produced by A.J.E. Child, president of Burns Foods Limited, on page 11 are very interesting—and indicate consumers may never see lower meat prices again. And the figures on page 3 given by R.E. English, Alberta department of agriculture statistician, should indicate the farmer has contributed very little to the rise in meat prices. Further insight on beef prices may be garnered from some Commons committee testimony on page 15. Also from Ottawa, testimony from another committee reveals how they were juggling wheat around during the dock strike. This is on page 13. And on page 16 Bob Shepp, a knowledgeable railroader, tells how the Canadian Pacific helps keep the grain pipeline to Vancouver filled.

Three widely varied, but significant, topics are to be found on pages 12, 17 and 4. The first tells how Dr. Travis Manning, University of Alberta economist, views the rise of country auction markets. The next is an outspoken view of private power by Percy Boulton of Farm Electric Services Ltd. And the last is how the Alberta department of agriculture was reorganized under the new deputy minister, Dr. E.E. Ballantyne.

One day while in the throes of producing copy for this report, Harry Isenstein from the apartment upstairs came wandering down and handed me a dozen hand-written sheets of paper, saying: "Maybe some of your readers would be interested in this story." I think they will. It's on page 5, the intriguing story of the Jewish colony at Rumsey. On page 6 is another fascinating story about a Chinese grocery man in Buffalo. I was particularly pleased with the way Helen Howe put together the story of Woo Sam with tenderness and affection. Jim Romahn is a young city hall beat reporter on The Kitchener-Waterloo Record, whom I met at the Royal Winter Fair. He should be a farm writer because he has an intense interest in livestock as his father had before him. His yarn about those high-priced Holsteins will captivate you.

URBAN–RURAL INFORMATION GAP
– *The Calgary Herald* – October 17, 1969

I have mentioned several times in recent months the information gap that now seems to exist between the urban dwellers and their rural cousins. Therefore it won't hurt to take another run at this subject, even if only to add heat and not much light to it. Frank R. Thomas is the man behind the inspiration to write. "I am often appalled at the sublime ignorance of city people concerning farming and the problems faced by farmers," he said. "I'm sure we are regarded as just a bunch of grumblers by a great many. However, it is just human nature in most cases to ignore anything that does not concern us personally."

Mr. Thomas knows both sides of the fence since he explains he has farmed since 1915 and still owns a farm east of the city, but has lived in Calgary for the past 12 years. He said he can cite many instances of misunderstanding. One of the most glaring cases of this sort to his mind was the rising price of beef in June when "everyone cried like stuck pigs. But nothing is heard now about the plight of the producer now that it has dropped 10 cents a pound on the hoof, regardless of whether that saving has been passed on to the consumer." Mr. Thomas said another concern of his is the grain market. It suffered last winter due to the inability of the railways to move grain to the Pacific Coast. "I'm sure we are still suffering loss of markets on this account," he contends. "One of the reasons for not getting the grain moved was the heavy snow and cold, particularly in the mountains. It is well known that the railways, especially the Canadian Pacific, were short of diesel power for moving grain trains. Previously they had been able to hire diesel engines from U.S. railways to help out but last winter they were unable to hire any. Apparently ice and snow getting up into the low-slung unit motors will cause them to burn out in a short time, hence the trouble. (At one time CP was reported to have 60 units out of service at its diesel base in Calgary.) "The point I am trying to make is: what are the railways and the government going to do about this situation?" Mr. Thomas asks.

He points out the Kaiser Coal Corporation has exerted pressure on CP to order many new diesel units in order to move coal to Japan. "Are the farm people of Western Canada not just as important to our economy? Do the railways plan to give priority to moving of coal and potash and lumber?" Mr. Thomas asks. This brings up another point on which he has devoted some thought: the Crow's Nest Pass rates! As have a great many other people, he speculates that perhaps the railways would be more interested in moving grain if the rate were raised—and the farmer would benefit in the long run by having his grain in the right place at the right time. "I know this idea is not popular with many farmers but why not face the facts and do something about it?" he suggests.

Incidentally, a paper on this subject was prepared for the Agricultural Economics Research Council of Canada by M.L. Lerohl, but he couldn't reach any definite conclusions because of the complex factors involved. U.S. railways, it is true, do receive higher rates than Canadian railways for moving grain, but Lerohl found many inefficiencies in their operation.

There is an exception to every generality. Not all city people misunderstand farmers. Eric Nicol, the wit who writes a column for *The Vancouver Province*, thinks Prairie farmers are nicer people than loggers and fishermen. That's why B.C. long-shoremen are loading farmers' wheat on boats and allowing lumber and fish to pile up.

NUMBER OF FARMERS DROPS, BUT PUBLICATIONS CLIMB — *The Calgary Herald* - October 16, 1984

A phenomenon of agribusiness is that as farmers become fewer, farm newsletters become more numerous. The latest publication is the *Canagrex* newsletter, which started up last month. It contains a raft of staff biographies and puffery on what *Canagrex* is going to do. *Canagrex* has done very little of note yet. The *Canagrex* publication has an uncanny resemblance to Canadexport, published by the Canada Department of External Affairs. This broadsheet is also jazzed up puffery for a government department. More uncanny still is that Vince Meechan, *Canagrex* newsletter editor, and Mike Gillespie, Canadexport editor, can be contacted through the zenith number of the Trade Information Centre in Ottawa. Is this a case of External Affairs swallowing up the Department of Agriculture's *Canagrex*, which blustered in on its territory when it wasn't particularly needed or wanted?

In fact, as an exporter of actual agricultural products grown by Canadian farmers, *Canagrex* may have been set up too late. According to a trade update in the rival *Canadexport*, one more nation which used to be a wheat importer has now become an exporter. This is Saudi Arabia, which has performed a modern-day "desert miracle." In 1975 Saudi farmers produced 300,000 tons of wheat. This year's crop is 1.3 million. Next year's may be larger. The Saudis can't eat all the wheat they now produce. The miracle has been performed with government assistance of a $22-billion investment in agriculture since 1980. This must have been a severe culture shock to the nomadic Bedouins, who comprise a quarter of the population of Saudi Arabia. Vast stretches of sand and scrub—not unlike the poorest land in the Palliser Triangle—have now been irrigated by drilling deep wells that are spraying water through pivot irrigation systems. Wheat yields have been as high as 85 bushels to the acre.

The Saudi government had the good sense to import important development tools at a cost of $1.4 billion a year. Professional farm management teams and top-grade livestock were brought in from highly developed agricultural countries. Also imported were the latest available machinery and technology. Massey-Ferguson Ltd. of Toronto got in on this machinery export bonanza and this helped put its books in the black again. Saudi Arabia has been able to cut its food imports by $1.3 billion a year.

Faced with the rising cost of food imports, other countries have gone the Saudi route to a large extent. With such startling new developments, Vol. 1 No. 1 of the *Canagrex* newsletter may become, a collector's item. There may be no more—as there may be no more Canagrex. Possibly the only thing that can save Canagrex is having Dave Durksen on staff. He was an information officer with old Federal Grain Ltd. and later went to the Federal Department of Agriculture in Ottawa.

A portent of the future may be seen in the fact that Peter Phillips of Agricultural Information Services of Exeter, Ontario, has been appointed North American advertising representative of *Arab Agriculture*. To be printed in Arabic and English, Vol. 1 No. 1 of the new publication will come out December 1st with a press run of 9,500.

In a recent column, mention was made of three new farm magazines for women: two in Ontario and one in the United States. Now there is another in Ontario. The weekly *Western Ontario Farmer* of London (which also publishes the *Eastern Ontario Farmer*) has assigned one of its farm writers, Agnes Bongors of Newmarket, to edit a monthly supplement, *Ontario Farm Women*. Peter Hohenadel of Milton, editor of the *Eastern Farm Journalist*, claims Canadian farm women haven't received such attention since the honeymoon. This brings up the question of a men's section in more established farm magazines. Well, *Farm Journalist*, reacting to demand, ran a story on how bachelors can entice women to their $500,000-to-$1-million farming and ranching enterprises. If they marry them, they become instant partners in such enterprises. The main thing to be able to supply to a date from the city is a reason why it is more important to do the combining than to take her to Banff.

The *Farm Journal* Ithaca, New York, is now running a lonely hearts club. This is reminiscent of the campaign carried out on behalf of rural bachelors by Ron McCullough when he was a farm broadcaster at CFAC Calgary in the early 1950s. He has never revealed his success rate for getting women married to farmers, however. Another zany U.S. farm magazine is *Farm Show*, which exists without ads. It recently appointed a new editor, Mark Newhall, a crazy farm writer who ran a recipe on how to make an automatic flapjack maker that produces 60 pancakes every three minutes. He also described an ironing board which doubles as a chair.

Also on a lighter note, the Alberta delegation to the annual meeting of the Canadian Farm Writers Federation in Winnipeg last month expressed the conviction this province is ripe for a weekly farm paper. This conclusion was drawn after hearing Tom Kent, chairman of the 1981 Royal Commission into the Newspaper Industry, gave his version of what is happening in the Thomson and Southam companies, which got their start publishing profitable newspapers. He alleges they are shipping cash flow out of their newspapers into other corporate ventures. Kent liberally handed out the further advice that to achieve greater profits, the papers are paring reporting staffs, overworking those left and pruning travel budgets. "Journalism is being downgraded," he bellowed. Thus fortified with a zeal to upgrade farm writing at least, the Alberta farm writers vowed to start a paper called Waco (Western Agricultural Co-op Organ).

Tom Kent, where are you now?

TRANSFORMATION OF FARM WRITING

As a distinct offshoot of the national press, farm writers historically found themselves short of money and praise as well as short of space to tell the agricultural story. However, being on the leading edge of everything new that was happening in Canada was usually sufficient compensation to the farm writers for the lack of some of these necessities. In addition to technological innovation, Schmidt argues out that government policies contributed to the decline in farm population and the resulting reduction in demand for agricultural commentators, nevertheless. He further argues for media independence to enable them to provide unbiased information to the public. Over the years, however, farm writing like farming has changed dramatically. Most of the farm writing today performs more of a promotional function for agriculture than reporting on various issues through an agricultural perspective. In the interest of fair play, and as a precaution against libel, Schmidt has never failed to offer rebuttal space in his columns for anyone who wishes to challenge his views.

NO PAY-TRON SAINT, PLEASE !
– *The Calgary Herald* – October 17, 1970

Most of the time the members of the Canadian Farm Writers Federation are a cynical hard-nosed bunch, not easily pushed around or persuaded against their will. I am quite sure the Canadian customs and excise think we are, collectively, a bunch of nuts or idiots or both. And they may be right in their assumption but the fact is our organization is almost broke. It's so nearly flat busted Jim Romahn, the hard-pressed secretary, hasn't got $65 to pay to one of the red-tape collectors who figures that publication of a list of members is worth that much. Storage charges are accumulating on the package in the Ottawa customs.

The way it happened was that we joined the American farm writers in publishing a list of members. The lists were printed in the U.S. and for several years came through duty-free. This time, however, the custom people broke their own precedent and, because the word "directories" was used to describe the contents, say the Canadian farm writers must hawk out $65. What they don't know is if we pay $65 to them, we will have to forego refreshments at the annual convention. The Toronto Press Club took us in off the street to even allow us to have a legal place to drink. In a spirit of true farm writer independence and defiance, Mr. Romahn has told the customs people the lists can "sit there till hell freezes over."

God save the Queen!

Over the years, farm writers have always put up a great show of self-sufficiency in the face of adversity. To my knowledge their show of independence has broken down only once. The occasion was the presentation of a brief to the Davey Special Senate Committee on Mass Media last April. The brief stated that the demise of many farm papers in the last several years has thrown many farm writers out of jobs. There can be no quarrel with this statement. However, the brief attempted to show that farmers need farm writers in order to stay solvent. Therefore, the brief concluded, the government ought to supply this need as one of its many services.

It seemed to me—and I may be writing a minority report on this—the brief didn't prove need. The reason the papers went out of business was that the farm population dropped from about 700,000 (at the end of the Second World War) to about 300,000. Because of the reduction in numbers circulation of farm papers also dropped (or became suspect) and advertisers couldn't see their way clear to buy space in them. Consequently many went belly-up. The drop in farm population is largely tied up with government policy. Federal and provincial governments have instituted a deliberate policy in the post-war years of forcing as many farmers as possible out of that vocation. Despite having the best technical information available, farmers could not hope to stay in business against the government policy juggernaut. The farm magazines were supplying this kind of information to farmers to the degree that they achieved a high degree of proficiency and efficiency.

It seemed to me, therefore, that what the farmers really needed in the way of coverage was a press that could tell them how to use government policy to make money. They were not looking for the technical stuff but valid information. Such being the case, how much would they get from any department of agriculture which undertook to supply the need for farm writers dishing out valid information they could use? The answer is plainly obvious.

The first time a government writer issued a release critical of government agricultural policy he would be sidelined, put in mothballs, sent to Coventry, or be put on the unemployment rolls. This isn't the only hazard to such a job. It is conceivable that if the government could foist an operation LIFT on farmers, it could also do the same to captive farm writers. Although I know several ministers of agriculture that feel like paying farm writers and other critics to shut up, it would be only too simple a matter for them to institute an Operation LIFT for their own employees.

One other point in the brief leaves a hole large enough to drive a horse and wagon through. No mention was made of the fact many farm organizations bankroll their own papers. Had the farmers really needed the services the brief claimed they needed, the farmers themselves through their own organizations would have seen to it their papers hired the displaced farm writers to carry on the services dropped when the privately-owned papers ceased publication. There was no organized move on the part of the rank-and-file farm organizations to beef up their own papers with these writers.

Now I ask you: Why should the government undertake a service the farmers have it in their own power to undertake—but haven't done it? Why should the government be called on to revive a dead horse? No, my minority report will have to be that once a farm writer dips into the government pork barrel, he becomes useless to the farmers he would serve. The department of agriculture has enough to do without becoming a pay-tron saint to farm writers, broke as we all are!

FARM WRITERS MAY HAVE TO SELL PRODUCT — *The Calgary Herald* - November 5, 1981

When farm writers get together their conversation often drifts to the age-old theme: In relation to the space and time given petroleum and energy industry issues by Alberta and Canada press, radio and television, why is agriculture deprived of the same priority?

The Calgary branch of the Alberta Institute of Agrologists wanted answers to this question too. It brought in Ted Byfield, publisher of the *Alberta Report* to discuss it at its annual meeting earlier this week. His perception of agriculture versus oil is that "whereas energy is vital, food production is more vital, essential and pivotal. Both are in a state of crisis." His magazine has a section devoted to agricultural issues and at times cover stories are written on agricultural politics. He praised *Herald* agricultural coverage as being "the best in the country" and admitted his editors pick up issues from the agricultural page and play them up. He finds the concept of an agricultural package is beneficial and practical for the reader and can't understand why most papers discard reams of good agricultural copy.

He kept tabs on the *Globe and Mail* for a week and found that its ratio of oil-to-agriculture stories was 69-to-4. Byfield suggested the oil business grabbed more ink and time because it had a better public relations effort. "Much of the agricultural commodity groups' public relations thrust is not aimed at the public but at politicians and other farm pressure and lobby groups," he charged. "This is a desperate mistake. You have to aim at the people who make the politicians act. Talk to the people who frighten the politicians. The way to get to the politicians is to get to the electorate. You have to learn to address your press releases to the most improbable target. Many of the farmers think that commodity advertising is 'nuts'," he said. "For instance, some of the milk advertising must curdle dairy farmers' stomachs. But it has an impact on the urban population—and sells milk."

One of the most terrifying prospects to the city reader is agriculture and its ramifications. But the average man-on-the-street in Calgary knows more about what Dome Petroleum is doing in the Beaufort Sea than he does about why the cattle business is in trouble at present. "This may be because the big cities made their most spectacular growth around the oil industry." It is Byfield's opinion the farmers and their organizations do too much pussyfooting around to get their story across. When Quebec farmers threw milk at Eugene Whelan, hammered C.D. Howe in Manitoba in the 1957 election and marched on the Alberta Legislature in 1976 to obtain a cattle subsidy, the people took an intense interest in agriculture.

There are controversial issues in agriculture but too often the groups involved become defensive about them. "A writer can't write about controversial subjects without hurting someone. However, if farm groups get into a fight that excites the public interest and the reader or viewer can get interested in the issue if it is well-handled," he said. "That will help tell the story."

In the eyes of this publisher the farmer is an amazing guy. The issues surrounding his business are absorbing. Many of them have large enough land holdings to make them millionaires on paper. Agriculture hasn't had a profound effect on disposable income the way petroleum products have since the politicians have grabbed the oil industry. If the farmer were catapulted into that position he may be accorded the dubious honor of getting more ink.

FOR A CHANGE, FARM WRITERS RATE PRETTY WELL — *The Calgary Herald* - December 19, 1981

During my time as a farm writer I have had little time to sit around wondering what people think of me and my writing. Had I done that, I might have turned into a spineless mass of jelly. As a result of this, there are, no doubt, a few persons who won't speak to me and probably a number who would like to get me in a dark alley. (Probably that's the reason the *Herald* has provided a group of us writers with what we jokingly refer to as bullet-proof glass.) However, I have never refused any reasonable reader whom I might have offended or criticized fairly or unfairly to have at me with all the space he wants for rebuttal. This has been a means of keeping most of the readers happy. The offer is good any day of the week. A few have availed themselves of the opportunity and debating points raised have made this a better column, I feel.

Members of the Eastern Canadian Farm Writers Association, probably because of their masochistic tendencies, recently gave the readers a chance to joust with them for all the bull they write. This occurred in a rare role reversal at a bull session during their annual meeting. They wanted to find out what's wrong with the performance of farm writers. The surprising thing that came out of this performance test by three newspaper readers was the farm writers are doing a pretty good job of getting the story of agriculture across to the general public. (Maybe the panel was afraid to let the writers have it between the eyes for fear of being eternally ignored in the press, a fate worse than death for any public figure. The panelists were Jim Chalmers, chairman of the Ontario Chicken Producers Marketing Board, Mrs. Ruth Jackson, vice-president of the Consumers Association of Canada, and Peter Hannam, past president of the Ontario Federation of Agriculture.)

Well, Chalmers confirmed he's not speaking to Kevin Cox, farm writer for *The Globe and Mail*. He wants the writers to take a broader view of the poultry industry; to lift the focus from the foibles of marketing boards and go into the news about feed millers, poultry processors, supermarkets and Col. Sanders and his finger-lickin' good chicken. This almost turned the panel discussion into a battle over: "What's wrong with supply management?"

Hannam said the farm writers were a bit one sided in that they tended to concentrate on the bad side of agriculture and an image is being created of farmers that may backfire some day. He wants to see the public more informed on the "good news" of technological advances and progress.

Mrs. Jackson said she has restored Jim Romahn of *The Kitchener-Waterloo Record* to her good books. She figures he has done a better job than most of putting agriculture into a broader context. She claims there is getting to be a larger, more dedicated and knowledgeable corps of farm writers across Canada, who are getting more and more space to get farm stories, farm problems and farm issues across. This gives farm issues more attention than other issues and the attention they get is unbalanced.

After this treatment, the farm writers could well sit around in a self-congratulatory mood and decide to do nothing more than they are already doing. In the case of *The Herald*, I think this would be wrong. The farm writers here are simply not being given enough space to tell the agriculture story as it is and I'll promise the readers to continue fighting for more.

CREAM OF THE PROFESSION

The Schmidt pen is active in praise of other farm writers whose work is outstanding in their field. In Schmidt's opinion Jim Romahn, of Kitchener, is the best in the country. He admires Romahn for using a bank scholarship to study biotechnology at the University of Guelph, years before biotech hit the headlines. Another is Don Baron of Regina, who, among his other accomplishments, wrote a book on the history of a divisive force within the grain growing industry of Western Canada.

THEY COULDN'T FORESEE TRENDS
- *The Calgary Herald* - December 6, 1974

My friend, Jim Romahn of New Dundee, Ontario, has retreated from the labyrinths of the Ottawa scene to go back to the *Kitchener-Waterloo Record* as a reporter. Romahn's most recent job in Ottawa was chief of the news section of the information services of the Canada department of agriculture. In this job, he dug out the facts for Agriculture Minister Eugene Whelan's speeches, which have put Mr. Whelan in solid with the farmers as an agriculture minister. What he has said, in essence, is that if farmers can't survive economically, then Canadians aren't going to eat as well or as cheaply as they have done. In getting this story across, Jim Romahn knows whereof he speaks because he came right from one of the best hog and dairy farms in Ontario.

Writing 170 speeches for Whelan was done at a price to his health and his doctor told him if he didn't get out of Ottawa be would suffer a heart attack or get himself a case of ulcers. So he went back to his former job on the reportorial staff at *The Record*, where the pace is hectic enough, but it must be a bit slower than Ottawa.

One of the first stories he came up with when he got back on the street in November was one that no doubt tickled any writer with a dairy farm in his background. This was the fact that margarine will soon cost more than butter. That story settled a lot of old scores around Kitchener and wherever dairy farmers get together to curse the ghost of the late Senator W.D. Euler of Kitchener for introducing his margarine legislation after the war. Up to the end of the Second World War margarine manufacture in Canada was outlawed to protect the dairy farmers. At that time at least 700,000 dairy farmers needed protection from cheap imported oils used to manufacture the product. None of the oils grown on Canadian farms then was suitable for making the product.

Senator Euler got his bill through the House of Commons at a weak point in dairy-farmer history. The government had dropped a subsidy program and a temporary shortage of butter developed. Not only did margarine come in but, the government then took the opportunity to put dairy farmers in a government straitjacket which still constricts them. Between the doctors condemning butter for its cholesterol content, the margarine lobby howling to be allowed to use the same coloring as butter and the government strictures, dairy farmers have had a difficult row to hoe in the last few years.

Thousands of cream producers were driven out of business by government interference and we now find ourselves in the position in Canada of having had to import butter for the last three years. One of the arguments the dairy farmers used against margarine and the lobby which wanted to adopt butter's color was that if the manufacturers could capture the table spread market with margarine (which was then selling for about half the price of butter) Canadians would later see margarine prices go to astronomical heights. The legislators, with their eyes on a cheap food policy for Canada, couldn't see that far ahead. But the dairy farmers were right. And if they are now doing a bit of gloating, they may be pardoned.

The retail price of butter in Calgary today is 94 cents a pound. The retail price of corn-oil margarine is $1.03. Some of the other vegetable-oil brands are 79 cents and 93 cents. The reason for increased margarine prices is that vegetable oil crops have been sharply reduced in yield in various parts of the world, but more particularly in the United States. It wouldn't surprise me in the least if dairy farmer organizations started using the same tactics against margarine as have been used against them by the margarine makers for years. Margarine men are now on the defensive. I take satisfaction in the current situation in that I have always insisted on butter on my table. I, too, have roots on a dairy farm. And for those margarine eaters who now envy the fact that I pay less, all I have to say is: "Pay up, you sods!" And to those who I know are going to switch to the cheaper butter, I say: "Welcome to the fold, but remember where your friends are."

AN ACE FARM WRITER ANSWERS HIGH CALLING — *The Stettler Independent* - November 9, 1994

Usually at the annual meeting of the Alberta Farm Writers Association a quiet time is spent on giving mention-in-dispatches to the departed. This year attention was concentrated on Jim Romahn at the *Kitchener-Waterloo Record*, the top writer in the field and a modern-day legend. I am speaking in the past tense as he left the paper when he was given a death threat. He was told if he didn't accept a buy-out package he'd be stuck on a desk job—a fate worse than death. Therefore at age 51 he found salvation in the former and took a temporary position in the hierarchy of his church.

This happened after he and 20 other staffers were rounded up one day and herded into the newsroom and given 15 minutes to say good-byes, then given the bum's rush out the front door by security so nobody would pour beer in the mainframe of the computer. This wasn't the first time Romahn had been given a death threat. The first threat resulted from a series of columns he had done on a myriad of sins of the meat packing industry across Canada. Some readers thought he had come down too hard on the packers. So did the packers and he became the victim of some Mafia-style threats. Just recently I found out what had precipitated the series. He had a friend within the ranks of a Kitchener packing plant, which had built up a reputation on quality processed meats. A new president came in and savaged his friend who refused anything better than the best. The new president was soon out. He ended up in Edmonton.

Before going to *The Record*, Romahn wrote speeches for Agriculture Minister Eugene Whelan of Canada. Whelan was noted as a hard taskmaster and Romahn burned out after six years and quit. At the paper he set out to tell the readers everything he knew about flowers, fruit, fertilizer and farming, not to mention pigs and politics. His fame spread far beyond this provincial paper. So accomplished was he at digging up facts that out-of-province farm writers used to call him about current issues in their own provinces. Often his paper couldn't find enough space for all his copy.

When he won a $20,000 journalistic prize, he spent it by taking four months off to study biotechnology at the University of Guelph. He wrote up a lot of his new-found knowledge but the deskmen couldn't understand it and much of it never made it into the paper. This same subject is now filling the newspapers today—10 years later. In his swan-song column, Romahn expressed the opinion producers should divert dollars from advertising and promotion to research and development. He said it this way:

"Farmers will gain more from backing beef quality than spending so much on Olympic swimming champions like Mark Tewkesbury to sell their meat. Hog farmers will get a better return from money invested in development of management and nutrition packages to go with earlier weaning than they will from spending millions on advertising. Dairy farmers will gain more from staying on top of the heap in the world-wide competition for better genetics than battling the beer and soft-drink company giants for fractions of a percentage point-of-market share."

Over the years he grew disillusioned with marketing boards as they became more bureaucratic and constantly increased their selling levies on farmers. He had a long memory about such things as spending good money after bad—such as a $10,000 grant from the agriculture minister to Jersey milk producers to spend on exploring a market for trademarked milk. In the late 1960s, he recalled, the Ontario Milk Marketing Board won a series of court battles to disallow this practice. And he was happy to see Agriculture Minister Ralph Goodale of Canada "do what former agriculture minister Charlie Mayer should have done: remove Bob May and appoint Tony Tavares of Toronto as a director of the Canadian Chicken Marketing Board. May was judged guilty of violating quota production limits on his own farm."

Yep, these kinds of columns will be missed.

DON BARON WOULD FREE UP THE BREADBASKET — *The Hanna Herald* - December 17, 1997

Richard Allen, the hired gun who wrote up the Social Gospel for Mel Hurtig's Canadian Encyclopaedia, had a poor understanding of the movement. He said "although the Social Gospel is often categorized as an urban middle-class phenomenon, it did attract a few agrarian reformers."

A Regina farm writer has just published a book which attempts to set the record straight: that the Social Gospellers did a great deal more economic damage to grain growing in the prairie provinces than Allen has accounted for. The Regina author is Don Baron, one-time editor of the Country Guide, who ran the agriculture and resources department of CBC TV for six years and served a term as president of the Agricultural Institute of Canada. The book is titled: "Canada's Great Grain Robbery." It is the largely untold story of the Social Gospellers holy crusade against capitalism. Their crusade took hold shortly after settlers arrived on the Prairies and after 1900 began transforming the land into a new world-size breadbasket. Europe looked upon Winnipeg and its grain exchange as the world's greatest primary wheat-price-setting agency. The Winnipeg Grain Exchange came to be regarded as more important than the Chicago Board of Trade for pricing wheat. The grain growers had an open market in which private grain companies became exporters to the world market. Grain was priced on the Winnipeg Grain Exchange. Thus the early Western grain trade was based on the capitalist system the United States had opted for. This option did not sit well with a number of influential U.S. fundamentalist preachers who considered it a devilish system.

They decamped for Canada to create a new heaven on earth. In Canada the devils of capitalism could be easier exorcised. The Social Gospellers arrived a time when the pioneer "sodbusters" were fretting about abuses occasioned by practices of the railways and some of the grain companies. Out there on those raw stretches of prairie they felt a sense of powerlessness because they were not big enough individually to do anything about fluctuating grain prices. The Social Gospellers were right behind the farmers when they decided to discipline the private trade by organizing grower-owned grain companies and, later, co-ops. But a great American theologian Reinhold Niebuhr, pointed to the "fact that institutions built to combat perceived abuses can themselves become sources of injustice."

Baron alleges the Social Gospellers' theology was false when it attacked profits and the private grain trade. They wittingly or unwittingly "shackled the vastly rich prairie breadbasket from its infancy to the present day with a system which became a giant monopoly run by the federal government." The fact that he is an elder in Protestant denomination has not dissuaded him from taking a position many grain growers would regard as heretical: that the church should not have intruded into politics and the economy. Baron draws his inspiration from his own vast knowledge of the prairie grain trade: his former boss Mac Runciman, one-time president of United Grain Growers Limited (who fought the Social Gospellers on their own ground) and Dr. Paul Earl, a former policy researcher with UGG, who wrote his doctoral thesis for the University of Manitoba on the subject. It is entitled, "Rhetoric, Reality and Righteousness."

Living in a province whose population today is the same as 50 years ago, Baron realized the impact the Social Gospel upon it. In doing so, he has written a blueprint for leading the grain growers out of the wilderness. Will the politicians and farm leaders listen to his homily and start to make the required changes? The answer: There must be a large body of people committed to carrying out the change. Some of those who have attempted that change have been whacked with enormous fines and have gone to jail.

This isn't the heaven on earth the Social Gospellers promised them.

GOVERNMENT INFORMATION

Obtaining meaningful information from any source has its vagaries and the government is no exception. Some information is handed out gratuitously by hard working government scientists. Some is stymied by the urban reporters who head for the press gallery bar during debates about agricultural issues. Fortunately, some writers like Schmidt persist when agricultural issues are debated. Periodically, farm reporters are lucky to hear government officials like Premier Ralph of Alberta speak to them. He can be funnier than Charlie Farquharson.

BUGS KEEP HER BUSY
– *The Calgary Herald* – September 13, 1966

The Canada department of agriculture, one of Canada's largest publishers, last year distributed 1,250,000 copies of the 500 titles it carries. One of the people who sent out a goodly number of the department's publications is Miss Isobel Creelman, who is in charge of the pesticide technical information office at Central Experimental Farm, Ottawa. People have a vast curiosity about bugs and they write her hundreds of letters wanting her to identify every kind of crawly thing. Miss Creelman took over the job earlier this year upon the retirement of Graham MacNay. One of Mr. MacNay's most memorable publications was on bedbugs.

A native of Vineland, Ontario, Miss Creelman graduated from Queen's University, Kingston, and after two years on fruit insect investigation, she joined the bacteriology division of the old science service of the agriculture department. She transferred to the entomology division in 1952 and has worked on the Canadian insect pest survey until her new appointment this year.

PRESS GALLERY EMPTY
– *The Calgary Herald* – October 20, 1978

Telephones Minister Allan Warrack of Alberta was a bit nettled last March 10th in the legislature when he looked up and saw the press gallery empty during what he considered an important speech by Agriculture Minister Marvin Moore. "The press does not seem to care about agriculture," Warrack complained to the farmers in his constituency of Three Hills. He offered to send them copies of Moore's speech to judge for themselves whether the whole pack of press gallery reporters had goofed off unconscionably. Moore used the speech to debunk two myths the government said were being spread by the opposition.

One myth was created by Bob Clark, opposition leader, who claimed Alberta is losing farmers. The two ministers concede there are people leaving farms. But they point to the fact there had been a net gain in farmers. Moore says there were 57,310 farmers in this province in 1976. This is 4,100 more than 1971. Warrack apparently used a different set of figures to show there are 5,800 more farmers now. Moore was more impressed by another statistic he had at his disposal. This figure reveals that 44% of Alberta's farmers are less than 45 years of age. "And we've more than doubled the number of farmers under 35 years of age since 1971," said the agriculture minister.

This is a dramatic turn-around. One of the big worries of farm organizations a few years ago was the advancing age of farmers. The largest group was in the 59-year age bracket. This figure of Moore's indicates the $2 billion agriculture industry in this province is becoming more vigorous and there is a new vitality in the rural areas of Alberta. The minister has seen a large number of young fellows in his department resign to go into farming. That is significant in a time when farming has been regarded as a dead-end occupation. Warrack was glad that in the same speech Moore went after Grant Notley, NDP leader, who keeps trying to explain how great socialism is working for Saskatchewan farmers.

Moore ran through a long list of specific Alberta government farm programs which are not available to Saskatchewan farmers but which help Alberta farmers become established and stay in agriculture. The minister went on at length to describe the programs. One of them was quite unusual in its concept and deserves more than passing interest. This was the home study courses in which more then 3,200 farmers enrolled last January. "Beneficial educational tools are needed in our industry," said Moore. The courses covered rapeseed (canola) production, hog production and weed control over three years.

One of the most significant steps taken by the Alberta government was in the field of trade and tariffs. This area is a deadly one for legislature reporters. They usually repair to the nearest bar when a minister begins a discourse on it. But trade and tariffs are important and exciting to Alberta farmers. For years Canada's trade negotiators got away with trading off Western agriculture for Eastern industry, despite the fact agriculture exports are the lifeblood of Alberta farmers.

The way it went was: Alberta sheepmen would be just about ready to market the lamb crop, only to see the bottom drop out of the market when it was flooded with Australia and New Zealand lamb. Upon sending a delegation to Ottawa, the sheepmen would be told: "Sorry, but we can't do a thing for you. The lamb is coming in because Australia and New Zealand have agreed to take nuclear reactors, car parts or kitchen sinks in return for exporting lamb to Canadians."

For the first time, the Alberta government has been responsible for putting a stop to this kind of nonsense in the current Tokyo round of GATT negotiations in Geneva. Moore pointed out Premier Peter Lougheed had forwarded a brief to Prime Minister Trudeau in December of 1976 on agriculture's place in multilateral trade negotiations. This a brief was submitted on behalf of the four Western provinces. It was a brief that didn't make headlines in itself but did get on the front pages when Alberta cabinet ministers followed it up with trade trips to Moscow, Tokyo, Geneva, Washington, Winnipeg, Teheran and Ottawa.

"I think," said Moore, in his March 10 speech, "we've been effective in drawing to the attention to the government of Canada, for the first time, the concerns of the agricultural industry in Alberta. I want to tell you a little story, Mr. Speaker, about what happens when we first started looking in the government files and records as to what kind of representations were made by this government in the Dillon round of discussions in 1957-58, and in the Kennedy round of GATT negotiations in 1967. We looked high and low in my department. We looked elsewhere, and quite frankly, Mr. Speaker, we could find nothing. No representations were made, so far as I was able to determine, on behalf of the farmers of this province at that time with regard to international trade and tariff matters."

I thought that, with Agriculture Week being celebrated in Alberta next week, this would be a fitting time to bring some of the important work of the agriculture minister to the attention of the urban readers. A young fellow in his department, Ken Hughes at Lac la Biche, pointed out to a young urban audience last weekend that the Alberta farmer is a pretty important guy. In response to appeals from a world which likes eating better than anything else, Alberta's 57,310 farmers have gone out into their fields this year and planted the biggest acreage of the six major grains—wheat, oats, barley, flax, rye and rapeseed—in the history of this province. When the last drop of recoverable crude oil is squeezed from the much-vaunted Athabasca tar sands then maybe some of the urban people, who don't realize how much of their bread originates in Alberta wheat fields, will be able to figure a way of shipping wheat through Alberta Gas Trunk Lines, Hughes concluded.

CHEWING OVER THE GOVERNMENT FOOD QUIZ — *The Calgary Herald* - October 24, 1979

It is Wednesday, October 24th 1979. This is Agriculture Week in Alberta. The wife hustles in with the TV Times and says there's a government food quiz on tonight. She wants to answer the 20 questions printed in the TV Times, namely the Charlie Farquharson's questions. The do-it-yourself at home quiz was taped at the University of Calgary Theatre and features Don Harron, CBC radio personality and actor, as quizmaster and his alter ego, Charlie Farquharson, as comic relief. Incidentally, Farquharson looks a bit like John Andrew—and takes the same approach to food: that it should be eaten with gusto.

John Andrew, head of the communications branch of the Alberta Department of Agriculture has a great number of advisers at his disposal. They will tell you how to go looking for bargains in the supermarket. However, this is clearly against government policy. Department policy is based on paying the farmer cost plus a reasonable profit and supply management in some commodities. There are supposed to be no bargains under this system. The Alberta government runs one of the biggest supermarkets in the province: the Alberta Liquor Control Board stores. There are no bargains there either.

Others of Andrew's advisers say a two or three-ounce piece of steak should constitute an average serving. It is to straighten out some of these misconceptions about food that I have taken it upon myself to answer the food quiz that your wife is busily pitting her wits against from 7 to 8 p.m.

Question: Where in the supermarket will you frequently find bargains?
Answer: I never go into those places. I can't stand the body checking.

Question: Is it true that a can of peas marked Canada Fancy is more nutritious than the other grades?
Answer: Peas should never be labeled because a pea is a pea is a pea, especially if eaten off a knife.

Question: How are ingredients listed on a label?
Answer: Mostly in French.

Question: What share of the average food dollar goes to the farmer?
Answer: Too small a percentage.

Question: What is the Consumer Price Index?
Answer: A hoax, according the Charlie Gracey of the Canadian Cattlemen's Association.

Question: What percent of Canada's land is suitable for farming?
Answer: None, if it doesn't rain.

Question: What does the grocery product symbol mean?
Answer: Something suspicious.

Question: Considering taste and nutrition, which is the best choice of green beans?
Answer: Straight off the plant, with the field heat cooled out immediately after picking.

Question: Do potatoes and milk offer you a balanced meal?
Answer: Only if you throw away the potatoes and drink the milk laced with a little rum.

Question: How many dollars worth of food does the average household throw into the garbage annually?

Answer: This a throwaway question but the answer is that it just depends who's doing the cooking. If you own a pig you make money on the deal.

Question: According to Canada's Food Guide what comprises a serving of lean meat?

Answer: The size of the steak you eat depends on whether you have been chasing cattle all day.

Question: How many of your food dollars are spent away from home?

Answer: Just depends whether your wife has more expensive tastes than your girlfriend.

Question: Why haven't you answered twenty questions?

Answer: It's 4 a.m.

AGRICULTURE BEWARE OF "INFORMATION" OVERKILL — *The Calgary Herald* - September 8, 1986

At the conference of the Agricultural Institute of Canada in Saskatoon, Dr. Red Williams, well-known animal scientist at the university there, pointed out during the plenary session—The Age of Information—that the sheer volume is growing at a rapid rate. It is doubling every three to five years, placing a great challenge on the system. He proposed establishment of "information centres" which would assemble information from around the world and boil it down into a useful form.

Not satisfied with the printed word, governments and private companies have introduced new technologies and information handling. James Bonnen, professor of agriculture economics at Michigan State University, claimed the tremendous volume is "transforming" the ways farmers make business decisions. Donald J. Blackburn, professor of rural extension at the University of Guelph, dwelt on the means of transferring the massive body of knowledge to the farmers and urged greater co-ordination between researchers and the various agencies who advise farmers. There has got to be a certain amount of overkill in the amount of "information" produced. The 1,200 members attending the Saskatoon meeting presented 500 papers— a new record. I would suspect that a farmer who tried to systematically absorb the amount of information available to him today could spend all his time at it—and not have any time left for farming. Inherent in the sheer volume of information available is the suggestion of a speed-up. However, farmers who live with it know that nature is hard to speed up. What farmers really need is the ability to scientifically assess the knowledge explosion and instantly exclude the trivia, chaff, nonsense, non-applicability and duplication. It is an abuse of the intellect to confuse a mass of facts with the ability to apply helpful material to a farm operation profitably.

Red Williams touched on this problem. Scientific information farmers already possess benefits them the least of all. They have no way of passing along the costs to consumers.

PREMIER RALPH CAME BACK TO THE REAL WORLD — *The Stettler Independent* - October 5, 1994

Premier Ralph Klein came back to his own September 15, 1994 when he was guest speaker at the annual meeting of the Alberta Farm Writers Association in Edmonton. He was the first premier who ever accepted an invitation to speak to this small but gung-ho group. He was accompanied by one of his powerful cabinet ministers, Walter Paszkowski. Before it became "an outrageous rock station," the Premier used to go on the air nightly at 10 o'clock to read "the Alberta Wheat Pool News" on CFCN, Calgary. "Yes," he related, "I would come on and say 'this is wheat pool broadcast No. 38,886'. God help you if you forgot the number and got it wrong. You'd get calls from farmers who kept track of that number." Those were the days of hard news on what was actually happening. Most of today's news is based on what some minority group thinks should be happening. At the end of the news broadcast the announcer extended birthday wishes to listeners and paid tribute to old-timers who had passed away. Then gravelly-voiced George Winkelaar came on with the livestock prices. The program ended with Rev. Ernest Manning, the Premier of Alberta, coming on with a few words of spiritual inspiration.

Premier Ralph made a point of noting that none of the big city media chose to attend the farm writers' dinner, although two television stations had specific invitations both from him and Bard Haddrell, farm writer president and head of communications for the Alberta agriculture department. The invitations were issued as the result of a "scrum" in hotel lobby before the dinner. He was asked to justify his "sin" of appointing Diane Mirosh, minister without portfolio and a critic of government social engineering, to the new Science and Research Authority at a salary increase of $25,000 during a period of rigorous government cost-cutting. The scrum, to the delight of onlookers, turned into a slanging match, with Klein unapologetically doing some top-rated media-bashing. When he got up to speak, he ad-libbed about 25 minutes onto the end of a 10-minute prepared script. He was somewhat steamed about the exchange with the TV reporters. He contended they were denying their audiences news of what was happening in the "real world."

"I told them if they wanted to know what's going on in the real world to come and rub shoulders with some farm writers and their agri-business guests. They looked down their noses and declined. They couldn't find any victims to get on tape in such a meeting," he said. He pointed out that the buzz word, "hi-tech," is highly important to this province (and that's why the government created the Science and Research Authority) but agriculture is just as important. It is essential to the continued well-being of Alberta.

Governments which continue to look after agriculture will continue to generate new revenue year after year because the opportunities for value-added production. "Many of the most capable MLAs are farmers," he added. "But few of these Legislature press gallery reporters will venture beyond the city limits when I go into the rural constituencies to discuss agrarian questions," he complained. "Most of the city media have dropped their full-time farm writers who know what's going on there." One example of the support the government is giving to agriculture and agri-business is to send a group led by Carol Hayley, MLA for Three Hills-Airdrie, into the rural areas to find out what government regulations stymie farmers. We will make recommendations for getting rid of the paper blizzard and outdated rules, laws and regulations; also to make the beneficial laws more effective.

The prepared speech was a report card on how the government is proceeding with digging itself out of its fiscal hole. "We are not apologizing for the fact this government has become the toughest political jurisdiction in North America," he said, "and I think taxpayers will come to appreciate our approach. We don't have a revenue problem. Our job is to get spending under control. We should soon be able to operate within our means, thank God."

When it came to thanking the guest speaker, the farm writers picked Bill Owen of Westlock, a farm broadcaster from Klein's time on the air. He had prepared a report on the premier's performance, as follows: "Based on a score of 10, we give you a 7 for debt reduction, 9 for media bashing, 10 out of 10 for adlibbing, 2 for reading the Wheat Pool News and 10 for being here." Owen later moved a resolution that Premier Ralph be given an honorary membership in the association. The motion passed, thus making him better equipped for the media scrums—as the Alberta Farm Writers Association never argues about the Constitution when they should be drinking beer!

AGRICULTURAL EDUCATION

As an author of several books himself, Schmidt has observed the Canadian Book Publishers' straightjacket, but he saw it torn open by a farm writer, Peter Lewington, who wrote a book on the history of the Holstein breed of cattle and persuaded a book publisher to print 24 deluxe numbered copies. Sold by auction, these copies grossed $24,000. Margaret Atwood, please copy.

The sophistication of farming brings sweeping changes in food production that provide exciting new career possibilities for those with the education to take advantage of them. Schmidt also notes the many other ways in which farm writers educated the public. For example, those civil servants who use weasel words or who take inaccurate public opinion polls, or indolent government press gallery reporters and newspapers covering the wrong wars.

JOB SURVEY – *The Calgary Herald* - September 8, 1986

Anyone who has a degree in agricultural economics or business administration is going to find that degree worth money in the next two to five years. A study run by the Agricultural Institute of Canada, funded with $115,000 by the Canada Department of Employment and Immigration shows professionals with these two degrees will be in high demand. The study was one of the initiatives resulting from the implementation of a human resource planning agreement negotiated between the department and the AIC in 1983. Cam Sibbald, of Calgary and AIC past president, said the survey went into companies' hiring plans and studied supply and demand for the next 10 years. Plant scientists, horticulturists, agricultural engineers and animal scientists will find job hunting easier, too.

WRITERS DON HAIR SHIRT
– *The Calgary Herald* - October 8, 1969

At its annual conference recently at Olds, the Alberta department of agriculture staff adopted the hair shirt approach. They invited in a number of farm writers, dirt farmers, educationists and researchers to ask these people what they expected of the department and what future action the department should be gearing up for. A session like this lends itself to little praise but much damnation. I think the department people took a fair amount of abuse with good grace in the sessions I attended. Fortunately there was plenty of constructive criticism. I don't know whose idea this was but it was an effective one. It enabled the staff employed by the department to feel the pulse of the taxpayers they serve and to either justify or re-evaluate policies, which they had put into force over the years.

In a way, an agricultural civil servant has a thankless job. I felt a twinge of sympathy for a great many of the old-time professionals on the staff. As the seminars went on, it became apparent that some of the skills which they had spent years acquiring had outlived their usefulness. This is the story of agriculture which has altered radically in the post-war years. The majority of the old timers had been trained to go out into the farming regions and teach farmers how to produce more of everything. They have done an excellent job in their field and the need for their particular services has now become somewhat limited. Farmers know how to produce and produce well.

The big need now is for advisers skilled in the art of farm management, marketing, social work, and food technology. How does one skilled in, say, the production of field crops (a skill which is almost passé in these days of large surpluses) expertly advise farmers on better management (skills badly needed and which require the skill of persons trained in chartered accountancy or economics)? These two skills have no real compatibility.

There is a body of thought that to justify the tax dollars which the minister of agriculture must ask of his constituents, the persons he has working for him must do a public-relations-image-building job for agriculture and farmers in the urban areas. There is a large job for the farmers to be done in the large urban centres. Urban folk have very little understanding of the important job agriculture does for them and for the nation as a whole.

As Don Robertson, a Carstairs farmer, pointed out "urban people have no idea of the cost, effort, or complexities of food production. They become a hostile element in the larger society against the farmers." Farmers, it is true, should be doing more missionary work for their own cause in the wider society. At the same time the professional agriculturists in the department of agriculture have a duty to make themselves known and make their profession known and how they are using it to try to feed the nation, in my view.

One of the problems the department will have to solve is how to accomplish this "infiltration" of the wider society. It will likely require a new kind of cat who can groove with the young swingers in the cities rather than trying to assist square old farmers grow more surpluses. Whether the department of agriculture will try the Olds experiment again for a few years is open to debate. It may take them a couple of years of licking to heal the wounds from such an experience.

However, if there is any consolation in it for them, the Eastern Canadian

Farm Writers Association used the same hair-shirt technique and came out pretty well scratched up. They called in a panel of farmers to tell them how not to do their jobs. One thing the farmers told the writers was that they took themselves too seriously. Too many farm writers think they hold the salvation of agriculture in their hands at a time when many of them don't even know if they're going to hold another pay cheque in their hands. What the farmers want to see in farm writers are those who keep the pot stirred up constantly. That's what they like. Stir things up but do it in an entertaining manner.

The two farmer-panelists said they hate to see farm writers quoting experts, especially if the experts have nothing better to do than tell farmers how to run their business. They also hate reading about Joe Farmer down the road who's doing such a fabulous job. Instead, they want to know why some man went broke. They think they can learn more from mistakes than from sermons. They don't want to read what's right about new policies, ideas and programs (these are all originated by people who can argue convincingly they are right). They want to read some intelligent criticism which will tell them what's wrong and where the weaknesses are.

My grapevine tells me the farm writers—as understandably as the department of agriculture staff—didn't take kindly to that kind of talk from their readers. It's really a great deal harder to find out what's wrong with new ideas than to simply quote the experts. It's a great deal harder for a writer to come up with bright copy than it is to make a few deletions on a press release or a research report, and paste it up. This is especially so when farm writers think they are the spreaders of the gospel that will either doom or save Canadian agriculture. However, it's great to have this dialogue going between the different segments of agriculture.

REASON FOR BAD WRITING
– *The Calgary Herald* – April 18, 1970

Words don't have meanings. Only people do.

The way different people sling words around creates problems for the people at whom the words are directed (provided they are directed at anyone). It is no new revelation that scientists have difficulty communicating with the man on the street, says Professor J.M. Ziman, physicist at Bristol (England) University. It is a little more difficult to believe that scientists have difficulty communicating with each other by means of the written word. But Professor Wolfgang M. Schultz, agricultural economist with the University of Alberta, makes this point.

People in the East have indicated that they don't grab all the words which have been scattergunned in the great grain talk-in on the Prairies. Some of them have been left far behind by government and farm politicians. Therefore, for the benefit of those who seek the wisdom of the ages on this great issue from the mouths of the leaders, here's the way to assess their words. (Conceptual reference is courtesy *The Farm Journalist* and *Punch Magazine*, which means I just stole their ideas word for word):

Premier Ross Thatcher of Saskatchewan says: "Saskatchewan farmers cannot be fooled."
He means: "Of course they can. But I'm not letting Lang get away with it."

John Diefenbaker says: "We want to get the government off the farmers' backs."
He means: "We want this government off the farmers' backs."

Charles Munro, President of the Canadian Federation of Agriculture, says: "On the other hand. . ."
He means: "CFA politics is a game with two sides and a fence. I'm staying on it."

Hon. Otto Lang, Minister in Charge of the Canadian Wheat Board, says: "Let's look at the facts about our grain surplus."
He means: "I'm losing this argument, so I'm going to hide behind obtuse statistics."

Alvin Hamilton, former Agricultural Minister, says: "I speak for the wheat grower."
He means: "I'll do anything for the wheat grower except become one."

Prime Minister Pierre Trudeau says: "We will build a new Western Canada."
He means: "Everything will stay the same but we'll send Otto and Bud west more often."

Roy Atkinson, President of the NFU, says: "The honourable gentleman is mistaken."
He means: "Honourable, my foot. The man's a blooming liar."

W.C. McNamara, Chief Commissioner of the Canadian Wheat Board, says: "We must meet the challenge of the '70s."
He means: "Can we please forget the hash that was made of the late '60s?"

E.K. (Ted) Turner, Chairman of the Saskatchewan Wheat Pool, says: "Saskatchewan farmers will not stand for this."
He means: "I'll send a telegram to Ottawa."

Agriculture Minister Bud Olson says: "The gentlemen in the opposition are trying to make political capital out of this issue."
He means: "They're on to a good thing. I wish I didn't have to defend that stupid remark Trudeau made in Winnipeg."

Opposition Leader Robert Stanfield says: "These are difficult questions and there are no simple answers."
He means: "What is he talking about?"

NDP leader Tommy Douglas says: "Nine wasted years in grain marketing."
He means: "The years I've had to spend on the opposition back benches."

Prof. Schultz makes a further point in this regard. He says true scientists—even in this MacLuhanesque age of television—should never neglect the written word. He says the failing of agricultural economists is they prefer to rely on word-of-mouth exchange for their information rather than the written word. What they do, in effect, is build a profession based on folklore rather than a well-documented body of reference material. He wants to see a greater effort made by the Canadian Agricultural Economics Association to stay on top of the information explosion by getting more of it into print.

FARM COLUMNISTS ARE EDUCATORS — *The Calgary Herald* - December 8, 1972

There was a great deal of glee recently when *Editor and Publisher*, a newspaper trade publication published in New York, came out with a story purportedly showing newspaper columnists ranked 16th of 20 in terms of honesty among selected vocations. The story once again proves my theorem the average newspaper reporter is not capable of reading a scientific report, deciding whether it will stand up and writing an accurate story. He invariably tries to sensationalize a scientist's findings.

Editor and Publisher reported the findings of Julian B. Rotten, a University of Connecticut psychologist, and Donald K. Stein, another psychologist from the University of Southern Florida. The first mistake *Editor and Publisher* made was in a name. It is Julian B. Rotter—not Rotten.

Phew!

These two psychologists undertook a study they entitled Public Attitudes Toward The Trustworthiness, Competence and Altruism of 20 Selected Occupations. *Editor and Publisher* didn't bother reporting a significant fact found in the first paragraph of the study—a fact that made the title a bit of a misnomer. The fact was that no cross-section of the public was interviewed. The subjects were 200 University of Connecticut psychology freshmen and sophomores, 50 school secretaries whose average age is thirty-eight, 50 school teachers whose average age is 35 in small towns in Connecticut and 96 University of Maryland psychology freshmen and sophomores. It is this sampling of predominantly white middle-class males and females cloistered in a small corner of the U.S. that Rotter and Stein attempt to pass off as "the public."

Phew!

Their ratings of 20 occupations selected by Rotter and Stein in terms of trustworthiness, competence and altruism were: 1. physicians; 2. clergymen; 3. dentists; 4. judges; 5. psychologists (how did they get in there?); 6. college professors; 7. psychiatrists (how did they get thrown in with the good apples?); 8. high-school teachers; 9. lawyers; 10. policemen; 11. TV news reporters; 12. plumbers; 13. business executives; 14. U.S. army generals (why were non-civilians picked?); 15. TV repairmen; 16. newspaper columnists; 17. auto repairmen; 18. labor union officials; 19. politicians; and 20. used-car salesmen.

It would be natural for a group of people with predominantly academic and pedagogical backgrounds to rate the people in whom they had the most confidence to be those requiring a university degree to enter an occupation. I sent for a copy of the Rotter-Stein report. For me, its credibility ended at the paragraph which began: "Knowing the low opinion most people have of used-car salesmen, this occupation was used as a kind of anchor group." I'm strictly from Missouri when a couple of shrinks try to choke a value judgment down my throat; it's comparing apples to oranges.

The most important thing *Editor and Publisher* overlooked was the Rotter-Stein terms of reference or motivation. Their terms of reference revolved around determining whom one can trust as society grows more complex. They were interested in finding out if the general public trusts "powerful" people in society.

They interviewed a group of people, the majority of whom are being educated or have been educated not to trust the political leaders, labor union officials, business executives, army generals and people controlling the communications media. And like puppets, their subjects came up with the "correct" answers, i.e., the political system created by these people cannot be trusted.

The Rotter-Stein conclusion was that if this group of leaders is going to keep the confidence of the public and reduce dissention and protest, they are going to have to quit lying consistently. Which is much akin to: "When are you going to stop beating your wife?"

Phew!

Too many scientists lose sight of the fact their principle role in a university is that of educators. A search of their pedigrees reveals many are using their tenure as a platform for half-baked political theories. They're usurping their roles as educators to such columnists as Dick Orr of *The Chicago Tribune*. Orr had his portrait hung in the Saddle and Sirloin Club at the Chicago International Livestock Exposition last month—a sort of an agricultural hall of fame. Agricultural Secretary Earl Butz of the United States pointed out at a testimonial dinner how Orr survived as a farm writer on a metropolitan daily. Most of all, as a farm writer, he is an educator—telling farmers the direction their own organizations are going but, more importantly today, interpreting trends in agriculture for urban people.

Despite the fact Orr's title has been changed to rural affairs editor, Butz pointed out the Chicago paper's management realizes rural development is catching on and public concern for the countryside is mounting. More and more readers are attracted to news about the rural areas. "Farm matters may ultimately enjoy a wider public exposure because columnists like Dick Orr have adapted and have won a broader audience," Butz said. "He keeps probing till he gets the explanation. When his story gets into *The Tribune*, it's going to be right, and understandable. When the farmer and the commuter alike read it, they know they have gotten the straight story and enough facts to make a sound judgment."

OTTAWA'S FARM WRITERS DON'T FOLLOW THE PACK
— *The Calgary Herald* - April 29, 1983

Farm writers tend to look down their collective noses at reporters in the parliamentary and legislative press galleries across Canada. Most of the gallery reporters are content to cover "the quick hits" and "hot spots" and little else off the beaten track. This opinion was expressed by Barry Wilson, Ottawa bureau chief for *The Western Producer* of Saskatoon, at the annual meeting of the Eastern Canadian Farm Writers Association. The press gallery reporters dog the politicians and lobbyists too much and don't do enough background digging and reporting. "They allow their contacts to set their agendas for them," he added. Wilson doubts the Freedom of Information Act, if and when it is ever passed, will have any impact on parliamentary coverage as "99% of the press gallery reporters don't read the reports they now get." Wilson is one of two full-time farm writers on The Hill. George Price of the Canadian Broadcasting Corporation is the other. Canadian Press has a staffer who is allowed to do part-time farm writing. Don Mazankowski (PC—Vegreville) said the farm writers who work out of the gallery do a good job but "there are not enough of them." Coverage of the important issues of farming and agricultural politics and Ottawa meetings concerning agriculture plus sessions of the Commons agriculture committee are given little coverage, he complained. Mazankowski also complained the Tories have gained no attention for their complaints that taxes have increased energy costs by 40%, including the cost of natural gas to make nitrogen fertilizer, and oil and diesel fuel to run farm tractors.

Anthony Westell, head of post-graduate journalism at Carleton University, commented the media are paying too much attention to isolated and sensational daily events and not enough to broad trends, so society has no background against which to understand daily news. Mazankowski agreed with this and said Canada's daily papers are ignoring important issues in the nation's food-producing industry. The readers are given pages and pages of recipes but skimpy space on how the commodities get to the would-be cooks.

AUCTION BY FARM WRITER – *The Calgary Herald* – January 7, 1984

Over the years Canada's authors and book publishers have written their best prose about how little money there is for them in their craft. They have tried all sorts of gimmicks to induce the public to buy books—from having them condemned by ministers of the gospel and ministers of state, to appearing on controversial radio talk shows, to filling them full of four-letter words. Book publishers have beat the bushes for government subsidies, authors have organized into sad little writers' unions and they have tried to change the copyright laws to force library subscribers to pay reading fees. But they have never been able to extract any more money than the public decided it wanted to expend for their product.

A product, you say? Let's take a look at what a book really is in the eyes of Peter Lewington, an Ontario farm writer from Ilderton. Sitting back in his farm home, he took a cue from the late Senator Harry Hays. Like Hays, he bred, raised and milked Holstein cows. Hays used to proclaim that the breeding of fine cattle is the work of an artist akin to a person with a facile pen, spreading oil paints or dancing in a corps de ballet. Lewington accepted Hays' definition of a Holstein breeder and an author. But he noticed one other thing Hays did with the product into which he put so much of himself. He used the auction method to sell the cream of the crop (accompanied by considerable ballyhoo).

I have written about Lewington previously in this column. The last mention of him was November 2nd when I wrote a review of his book, *Canada's Holsteins*, published concomitantly with the 100th anniversary of the Holstein-Friesian Association of Canada. I said all the right things a book reviewer is expected to say—but still there is more to the product than a review by another writer. One thing I omitted is the fact book publishers can be relatively stupid. Lewington received rejection slips for 2 1/2 years but overcame one from Fitzhenry and Whiteside by persistent lobbying.

Remembering that Holsteins are sold by auction, he decided to show the book trade a wrinkle it had never thought of to create demand and to extract the correct amount of money out of the public. He went to Fitzhenry and Whiteside and asked for 20 deluxe numbered editions. He then offered them at three different purebred Holstein auctions on the theory that if breeders would pay $1,200 for a good cow they would pay that much for a book about their favourite breed. It was sort of a vanity thing but he was right. Those 20 books sold for $24,000 before the regular run was even in the bookstalls—for an average of $1,200.

"The response was a milestone in publishing and probably a candidate for the Guiness Book of Records," he told me. Lewington received a boost from some of his old cronies in the breed. The first five were sold November 14th at the Royal Round-Up sale at the Brubacher Sales Arena in Guelph. Allison Fawcett of Winchester, president of the Holstein-Friesian Association of Canada, read the book's pedigree and it was auctioned for $6,000 to S.B. Roman. Roman is a breeder of top Holsteins at his Romandale Farm at Uniondale, since 1950, and is president of Denison Mines Ltd

Bidders showed up from all over Canada, the United States, the United Kingdom (from whence Lewington emigrated in 1947) and Italy. Ten were sold at the Royal Classic Holstein sale at the Royal Winter Fair, a sale originated by Hays. The pedigree was read by George Clemons, of Brantford, a secretary-manager of the Holstein Association for nearly half a century. The last five were sold at the Allangrove Triple Threat invitational sale at Brubacher's, with the pedigree read by Dr. Rusty McDonald, general manager of Western Ontario Breeders Inc. at Woodstock.

One of the successful bidders at $1,000 was Gordon F. MacLean, vice-president of Canadian Pacific Investments Ltd. of Calgary. Why him? CP Investments, through the Masterfeeds division of its Maple Leaf Mills Ltd., sells a lot feed to purebred cattle breeders. And going back in history, Canadian Pacific Holstein herd at the Western Irrigation District headquarters farm at Strathmore won the third master breeder shield of the association in Canada in 1932. CP is closer to agriculture now than merely hauling grain at compensatory rates.

The book sale was a great coup for Lewington. He was intrigued to witness a scene at the Brubacher parking lot following the final sale. British Friesian breeders who had failed to get a copy of the limited edition tried to buy a copy from an Oxford County, Ontario, farmer. However, he refused to be tempted, despite the handsome profit which was offered. (The standard edition of 352 pages costs $40.)

Lewington describes his association with the Holstein breed as like having an international passport. They have been exported to 90 countries. "The Holstein is an international celebrity accepted right around the world," he said. One of its little-known accomplishments was to save the Swiss dairy industry from destruction. The main Swiss breed, the Fribourg, was facing genetic defects earlier this century. But frozen Holstein semen introduced into the Fribourgs "rescued the breed from oblivion."

REPORTERS NEGLECT PRAIRIE WAR
— *Alberta Farm & Ranch Magazine* – February 1992

I fail to understand why the Canadian news media are spending millions sending their war reporters to cover a war between the Croats and Slavs in far-off Bulgaria while ignoring a war of gigantic proportions in Canada, namely the Crow Benefit War. True, the Canadian war has not been fought for 1,000 years but it is almost as complicated as the Croat-Slav conflict.

Our war was touched off away back when the Saskatchewan Wheat Pool was merely a pothole on the Prairie landscape. At that time, Wes Ball, the secretary, came up with the concept of the pools being all things to all farmers. Every input required by farmers would be handled by the pools and every commodity sold would be handled by them. The financing, or backing for financing, for Ball's proposal would be provided by a large reserve fund garnered from handling grain in the system of country and terminal elevators they had built. Not all farmers and not all pool members agreed with is concept and they charged that with gigantic financial reserves the pools would be able to create havoc in many segments of agribusiness. This issue has been debated and fought out for 60 years. After simmering down in 1983, the fight has flared up anew at the present time. That skirmish concluded with the passage of the Western Grain Transportation Act of 1983.

The pools, in concert with Quebec farm organizations, had enough political clout to have the Crow benefit paid to the railways. Quebec was concerned that if the grain growers collected the Crow benefit (now worth $720 million annually) it would tend to reduce exports because the railways would abandon their branch lines. The huge grain stocks which would result would see the Prairies competing with Quebec in livestock and poultry production. The pools' rationale for paying the Crow benefit to the railways was to keep the money out of private hands in agribusiness. They wanted none of that, as their grain elevator revenue would shrink towards zero thus eliminating the Ball concept. That was supposed to be the end of it and everyone would live happily ever after.

It wasn't. Several commodity group coalitions were organized to fight this method of payment. There were reorganizations of the coalition and the newest sees the Agricultural Diversification Alliance carrying on the war. Somewhere along the way, the railways got the government to pull out of the agreement to retain a bunch of grain-dependent branch lines until the year 2000. Prairie Pools Inc. has had to fight off several setbacks. One is that its biggest ally, Quebec, is threatening to pull out of Canada. The other was a revolt in the ranks in 1987 when a telephone poll of active members of the Alberta Wheat Pool showed 62% favored paying the Crow Benefit to producers. However, the dissidents were yanked back into line and discipline returned to the ranks. Prairie Pools Inc. is now trying to discipline the Canadian Federation of Agriculture to tell its members to button up their lips and not support the Diversification Alliance and its militant arm, the Western Grain Transportation Coalition, in its battle to change the method of payment. Coalition chairman Larry Maguire of Elgin, Manitoba let off a couple of salvoes about this, one of them being that Prairie Pools' pressure was blackmail.

The Wes Ball concept is based upon growers delivering grain through pool elevators. In the last half-decade of low prices many farmers have resorted to the ancient practice of loading their own cars and thus saving hundreds of dollars a car. The grain industry has been startled by the fact nearly 10% of car loadings are now in producer cars. Running a bit scared over this trend, Prairie Pools Inc. began to lobby the Canadian Wheat Board to place severe limitations on grain growers' access to and use of such cars. United Grain Growers Limited let off a blockbuster by taking a page ad in Saskpool's own paper, *The*

Western Producer, to denounce this lobbying effort. It made public a letter to the CWB not to get involved in removing growers' historic right to load their own cars under the Canada Grain Act.

The only factor, which will allow Prairie Pools Inc. to mitigate this staggering blow (which goes right back to the reason for establishment of the United Grain Growers and the wheat pools), is Prairie Pools' contention that payment of the Crow Benefit to growers will see retaliation against Canada by the GATT agreement and its negotiators. Canadian representatives to GATT are putting pressure on United States and European Economic Market countries to end the world grain price war which has clobbered Canadian grain growers and which has been occasioned by huge production subsidies in those countries. The reply to Canada has been: "You are subsidizing your farmers to the same or greater extent as us. So go to hell."

These are the broad dimensions of the war on the Prairies which the war correspondents are neglecting.

FREEDOM OF THE PRESS

Schmidt found it odd when Agriculture Minister Jack Messer of Saskatchewan began collecting farms for the government land bank but objected to the Southam Company collecting newspapers. His observation shines a glaring light on an extremely important question that has profound implications for the future: What is the critical difference between a private monopoly and a government monopoly? Schmidt argues that the ownership of one share of a private company entitles shareholders to sufficient recognition to enable them to voice their opinion at the shareholders meetings. But, what of those without shares? The current understanding of government is that it has an obligation to all citizens and, to a lesser extent, anyone within its territory. Witnessing today's increasing divestiture of responsibilities by the nation-state raises questions about the responsibilities of private monopolies.

Schmidt also deemed it peculiar that the Saskatchewan Wheat Pool's Western Producer tried to keep vital information away from farmers. He was not sheepish in declaring it farcical that the federal government took Tom Kent, a one-time shepherd on its Devco payroll in Cape Breton, Nova Scotia, to head a Royal commission on concentration of newspaper ownership.

THREE HILLS FUA LOCAL FAILS TO MUZZLE PRESS — *The Calgary Herald* - December 19, 1961

On the final day of its convention, the executive of the Farmers Union of Alberta beat off a rump resolution from the Three Hills local to ban the press from its meetings and adopt the hand-out method of press releases. The executive was no doubt embarrassed at the resolution because this year it had organized a committee to assist the press in its coverage of the annual meeting. The executive invited several members of the Canadian Farm Writers Federation to discuss with it more effective ways and means of presenting the public with a better image of itself.

However, despite any embarrassment caused by the Three Hills local, the executive realized it was the democratic privilege of that local to bring such a resolution before the meeting. The members-at-large used their democratic right of the vote to defeat the resolution. It is to be assumed the majority thought the press was doing a good job, or that they didn't want to allow the muzzling of the press to fall into the hands of the delegates from Three Hills.

THEY'RE STILL ASKING QUESTIONS
– *The Calgary Herald* – March 5, 1973

Now I know why Agriculture Minister Jack Messer of Saskatchewan didn't take advantage of my offer to get him into the newspaper business. He went into the meat-packing business instead with the purchase of 45% of Intercontinental Packers Ltd. May we now expect to see Messer's meatballs at our favorite meat counter? I made my offer to the minister during a question-and-answer session at the annual meeting of the Palliser Wheat Growers Association in Regina in January.

I had requested permission to ask Mr. Messer a question which was bothering me. I had wanted to ask the minister this question ever since November 12th when he went to Toronto to speak to the annual meeting of the Canadian Farm Writers Federation. Before he got down to the libretto—which was to outline to the writers some of his startling agricultural policies—he couldn't resist the temptation to give the farm writers a lecture. It was a two-point lecture.

In point No.1 he expressed some concern about the gradual concentration of newspaper ownership into fewer and fewer hands. This is, of course, a matter of public record. This concern of Mr. Messer hit directly at me because I happen to own some shares in one of the newspaper publishing companies to which he referred. My ownership of shares in the Southam Press Limited came about through an estate. Ownership of the company is invested in thousands of hands across this country from guys like me to farmers, plumbers, widows, machine operators, legislators, doctors and teachers. An annual meeting is held every year and the company directors are held accountable to the shareholders.

I was naturally interested, therefore, in Mr. Messer's point No. 2. It was that the more power and responsibility accruing to a group or individual, the greater the tendency for that powerful group or individual to cop out in their several obligations and responsibilities to the people. I couldn't really understand Mr. Messer's criticism or concern about Southam collecting newspapers—especially not when he admitted to the farm writers he is in the business of collecting farms. He is collecting them with public funds. He said he may collect 1,600 this year for about $20 million. He has a head office executive committee outside of the department of agriculture doing this. And that committee is directly responsible to him.

Mr. Messer further stated he would not sell me, a resident of Alberta and a fifth-generation Canadian, one of the farms he plans to collect in Saskatchewan because I don't happen to live there. However, I didn't feel so parochial or kind towards Mr. Messer. I made him an offer to get him into the newspaper business for $30.50—which is the value of one share of the stock I own. This would give him enough ownership to be able to approach the company's annual general meeting if he didn't feel the company's newspapers were disseminating the conventional kind of Socialist wisdom he espouses. There are other means of approach, of course, but ownership can be important.

However, Mr. Messer did not accept my offer. And, as I said, the reason became apparent last week. I am not sure that he answered my question to my complete satisfaction. My question was that as Canada's largest and most powerful collector of farms, what assurance could he give the people of Saskatchewan he would not cop out on his several obligations and responsibilities to them—remembering that he had transferred all those millions of dollars from their pockets to his pocket?

I could not see Mr. Messer's face but I am told by some of the 500 who were in the audience that day that he turned white, his jaw dropped and he leaped to the microphone in a bit of a fury—and gave one of the best speeches in his life. He explained to the best of his ability the reasons why his government had decided to go into "the radical departure from traditional land transfer patterns." However, despite previous explanations and many public hearings around the province, the land bank concept drew a number of elemental questions which indicate many people do not understand what it is all about. Questions were received such as:

- What kind of a point system does a young farmer have to have to qualify for a land bank farm?
- What assurance has a father who sells his land to the land bank that his son will have priority should more than one applicant apply for the same land?
- You have spoken of maintaining the continuity of the family farm. How do you justify the imposition of succession duties on family farmers?
- Why not let a father transfer land to his son free and thus eliminate the need for a land bank?
- Is it the intention of the Saskatchewan government to enter into an agreement with the federal government on its small farm development program?
- Is it your desire to eventually put all farms and land in Saskatchewan into Crown ownership?

There are those who will be critical of Mr. Messer and his cabinet colleagues for getting into the meat-packing business. I am one who won't be critical, for a good reason. In matters of taxation, Canadians may have not had much to complain about up to now as high taxation has been accompanied by low food prices. However, now we have both high taxation and high food prices. The politicians, having larger platforms than the packer and the retailer, have convinced the taxpayers meat prices are too high in a clever ploy to mask the fact we are overtaxed.

His entry into the meat-packing business will give Mr. Messer the opportunity to reduce meat prices. As a meat packer, he thus has the opportunity of the century to restore the credibility of politicians everywhere!

PAPER CHARGED WITH SUPRESSION
— *The Calgary Herald* - December 5, 1973

The publishing subsidiary of the Saskatchewan Wheat Pool was charged with "trying to suppress communications to farmers" by an official of the Rapeseed Association of Canada. "It's thought control," said Ken Edie, Dugald, Manitoba, information director of the association. He said Bob Phillips, new editor and publisher of *The Western Producer* of Saskatoon, refused to run an ad for the association in this week's issue. The ad advised farmers to "vote for rapeseed being kept under the present open-market system."

The Saskatchewan Pool favors placing marketing and pricing of the crop under the Canadian Wheat Board. Mr. Edie said *The Western Producer* has lost all its credibility as a self-styled independent farm paper "and reveals itself as nothing more than the political arm of the Saskatchewan Wheat Pool." He said the pool has been playing dirty politics by pulling out of the rapeseed association it helped to establish and buying space on Alberta radio stations advising farmers to change the system of marketing. Mr. Edie was speaking at one of a series of information meetings sponsored by the Alberta department of agriculture and the Rapeseed Association of Canada.

Praise was given to the Alberta Wheat Pool for participating in the information meetings to debate against those favoring the free market. The Saskatchewan Pool refused to debate the issue at public meetings in that province. The ad which *The Western Producer* refused to run was headed: "Did you know you can vote for the open market system, yet still get a pooled price for your rapeseed?" Mr. Phillips refused to run the advertisement on the grounds it would "muddy the waters," said Mr. Edie. "I think it's a blatant attempt to prevent rapeseed growers from learning that if the present system is retained, the federal government has promised to allow farmer to opt for a pooled price..."

The Saskatchewan Wheat Pool while refusing the rapeseed association the right to advertise in its paper (which has 150,000 circulation in Alberta, Saskatchewan and Manitoba) has been spending thousands in a high-powered advertising, campaign. Most of the pool's advertising campaign has said the present pricing system is "rapeseed roulette" or gambling.

It has been learned the Alberta Wheat Pool delegates decided to spend over $5,000 on television advertising advising its members to change the marketing system. *The Western Producer* has lost $200,000 to $350,000 per annum for some years. The loss is picked up by the pool.

Mr. Edie drew the analogy that the paper's refusal to publish the rapeseed association's ad is comparable to a Conservative newspaper refusing to handle Liberal or NDP advertising.

SHEPHERD TOM'S THEORIES PROVE UNWORKMANLIKE — *The Calgary Herald* - October 5, 1982

Canadian daily papers feel they have been led like lambs to the slaughter by Tom Kent and the Royal Commission on Newspapers. Kent has put forward a number of unproven theories, which those in the trade believe will fleece the newspaper business if the government passes legislation now proposed as a result. The publisher of *The Calgary Herald* has not been sheepish in hammering the theories of Kent, a former Winnipeg newspaperman, pet lamb of Lester B. Pearson, and a one-time shepherd in Nova Scotia.

There is no accountability of the government's action in making him head of a royal commission investigating the newspaper business at a time that he hadn't yet made an accounting of the sheep fiasco when he was president of Devco, a Crown Corporation which was set up to spend federal and provincial taxpayers' dollars to attract speculative ventures to the Cape Breton area of Nova Scotia. Because his mother was a shepherdess in his native England, Kent hit upon a plan to revive the flagging sheep industry. Where once 100,000 sheep had populated Cape Breton, by 1972 that number was down to 2,700.

He drew up a plan whereby Cape Breton rack of lamb would be eaten by millions of diners in fancy restaurants across Canada every night. That was a fair objective for which lambs should be led to the slaughter—probably based on the Alberta slogan that 10,000 coyotes can't be wrong.

But when he went about implementing that blueprint he didn't hire an experienced shepherd like Buck (the Shepherd) Valli or Mme. Benoit. He hired—hold your laughter—an editorial writer named David Newton. When Newton got on the job, he reported back to Kent two things stood in the way of the grand plan. The first was that no matter how many sheep he counted at night that flock of 2,700 wouldn't provide enough racks of lamb to supply those restaurants on a continuity of supply basis. The second was there weren't enough fences to hold more than 2,700, nor keep the coyotes away from them.

It was necessary to implement a program to solve the last problem first. Kent thus set up a program of low-interest loans for sheep fence. Then he borrowed Alberta's plan to give subsidies to farmers or would-be farmers to retain ewe lambs for breeding. But still there weren't enough sheep. To overcome this deficiency, Kent undertook a spectacular breeding stock importation from Scotland. This scheme involved a tremendous amount of red tape and a tremendous amount of money because imported sheep are required to undergo a three-year quarantine period to ensure the imports are not infected with the dread disease, scrapie. North Country Cheviots were flown from Scotland in two drafts—1,500 in 1975 and 1,200 in 1976. In 1978 and 1979 the breeding stock and progeny were dispersed at large sales at Mabou, Nova Scotia.

But rather than win plaudits for its $1.7 million import and quarantine efforts, Devco now stands accused by sheepmen of introducing two exotic diseases into Canada in its importation—diseases not seen here before. They are abortion storm and pulmonary adenomatosis, a fatal lung disease. Both diseases were diagnosed in Devco flocks and in the flocks of sheepmen who bought Devco sheep. There were charges that Devco tried to pull the wool over people's eyes and keep the disease outbreak quiet. Agriculture Minister Eugene Whelan ran an internal investigation and said nobody could prove the incidence of the disease was caused by the negligence of federal health of animals veterinarians.

An *Atlantic Insight* magazine writer reported that although the introduction of the diseases may not be pinned on Devco, the incident has plunged the sheep industry into a "morass of mistrust, animosity and paranoia." There are charges that Kent and Devco brought too many inexperienced people into the industry, they have suffered losses and are using the diseases to try to get the government to slaughter the sheep and pay them compensation. There are now only 8,000 sheep on Cape Breton. Tom Kent is gone but he isn't chewing on rack of lamb in Ottawa. The ex-shepherd is engaged in chewing up the newspaper business and trying to put it on the rack.

ELECTRONIC MEDIA

Only three years before the birth of John T. Schmidt, Canada's Inspector of Radio, Donald Manson, issued the first broadcast licence to Canada Marconi in 1920. In 1936 CBC succeeded the CRBC radio operation established in 1932 and in 1952 launched into television broadcasting. Having started his career in the newspaper industry at about the same time gives Schmidt the unique advantage of having ob-

served the developments in the electronic media from its inception.

He remembers when the CBC began as a national network with a strong corps of farm writers. However, farm writers were sent out to pasture after Margaret Lyons declared the farm programs were ghetto broadcasting. Schmidt reports on the farmers banding together to avoid the CBC's neglect of agriculture, but the die was cast and much of the farm broadcasting succumbed to the same fate as befell many farmers.

Schmidt provides examples of how interest groups utilize the media for various purposes and, conversely, how media affects people, particularly politicians. Undoubtedly, the electronic media enables more information to be disseminated faster but whether this format promotes better decisions is a question for the future to judge.

ATKINS WENT WORLDWIDE
– *The Calgary Herald* – July 17, 1980

A fellowship in the Agricultural Institute of Canada (AIC) has been conferred upon one of Canada's best-known farm writers, George Atkins, senior agricultural commentator for the Canadian Broadcasting Corporation. The fellowship is awarded by the professionals employed in agriculture for "professional distinction worthy of national recognition." No better description could fit the work of Atkins for his role of telling the great story of agriculture to his CBC listeners over CBL, Toronto, for over 20 years. Although his broadcasts attracted high listener ratings, it is ironical that Atkins received recognition from everywhere but the CBC.

The super-sophisticates of the English radio network, preoccupied with their own ratings, have been nearly successful in an effort to weed out his kind of broadcasting in Ontario. The farm broadcasts have been watered down to consumer-oriented pap, in the opinion of a large segment of the farm community. A further irony of the situation is that, although Atkins has not broadcast on the CBC for the last three years, his farm-writing efforts are now heard by 91 million persons—more than the CBC ever dreamed would listen to such broadcasting. That has come about this way:

To make a long story short, Atkins interested Massey-Ferguson in undertaking a project for developing an international farm network. Farmers of the Third World, who comprise a majority of the population, depend upon grass-root level farm broadcasters—whom they know and trust—for a great deal of their knowledge of advanced farm technology. However, some of those broadcasters are limited in their scope and don't have the budgets to travel either at home or abroad. Atkins' idea of the developing countries farm radio network, which acts as a catalyst in transfer of appropriate agricultural technology that can be used in farm broadcasts from one Third World country to another, caught on pretty fast.

After scouring Third World sources such as farm broadcasters, agricultural scientists, teachers and specialists, he sits down and voices a series of nine tapes. Along with the tapes, copies of the scripts are enclosed, plus introductions and cues for the announcers, and sent to stations in 65 countries on the network. He is doing essentially the same job as he was doing in Canada—but on a worldwide scale. While Massey-Ferguson picks up the tab for this service, there is no hard or soft sell in the tapes by the company. It is at the discretion of the recipient Third World broadcasters whether the company name is mentioned in the introduction.

Atkins stresses the point that "while the initiative for the service comes from the developed side of the world, developing country recipients have full control of what they use and how they use it. This is appreciated by them as nothing related to the project is being imposed on them from the other side of the world." A high percentage of Third World farm broadcasters are also local agricultural extension workers who communicate with farmers in other ways, both face to face and through the written word. Thus, many are using the material from Atkins in a number of ways. Bureaucratic red tape and obstruction at both national and local levels are among the most formidable hurdles faced by a great many Third World development programs. Obstacles of this kind have been totally bypassed by the network and, in cases where government bureaucrats have learned of the service, they have fully supported it.

As a farmer in his own right (Atkins has a farm near Oakville), he understands the hopes and aspirations of farmers everywhere. This has helped make the worldwide network successful and well received.

ALLEN QUIT PLAYING GAMES
– *The Calgary Herald* – July 22, 1980

Ted Allen of Taber, Alberta belongs to the New Wave of young educated farmers who have learned how to use their land to produce an abundance of food. The success of this aggressive, hard-driving New Wave has been achieved by independent thinking which has begun to put agriculture and food production more and more on the front pages. But as important as agriculture is to Canada and the world, leaders of the New Wave have become increasingly angry and frustrated by the low priority given to agriculture's needs by Ottawa.

Allen experienced the frustration of seeing the low priority given agriculture on two national committees on which he has served. As first vice-president of United Grain Growers Limited, he was appointed UGG's official delegate to the Canadian Federation of Agriculture. He was also nominated to serve on the farm broadcasting advisory com-mittee to the Canadian Broadcasting Corporation when it was set up in November, 1978. He was amazed to find that the Energy Supply Allocation Board, a federal government body created in 1977, proposes to downgrade agriculture to the "essential" category from the "critical" list for allocation of petroleum supplies during an emergency.

There is a bit of a crunch on fuel supplies at present of which many Canadians are not aware. When Iran cut off Japan, other countries agreed to reduce their purchases to allow Japan to recoup its supply. For Canada this has meant a reduction of petroleum products at wholesale level. Agriculture is now in the critical priority, i.e., the allocation for health, welfare and security (firefighting, public transport and police). The board proposes to move agriculture down to essential, the same priority as garbage collection, car manufacturing, snow removal and forestry. This came about when some pipe-smoking member of the board said as long as gasoline is used by agriculture in the growing of tobacco, it couldn't be allowed to remain in the top category. "The allocation board has misread the importance of agriculture," said Allen. "The members have been taken in by the myth Canada has tremendous surpluses of food. Food is the most essential need of the population. The board does not realize most crops are perishable and must be harvested when mature."

As a member of the federation executive, Allen will stay and fight this one out. In the case of the CBC watchdog committee, he packed his bags and resigned in March after he discovered "the CBC had more people of independent mind on this committee than it wanted. The CBC hadn't been honest with us. I had better things to do with my time than play games with them."

Al Johnson, president, was pressured into setting up the 18-person committee following criticism heard by the House of Commons broadcasting committee that the CBC was watering down its farm broadcasting. Only 13 were appointed, despite a promise by Peter Meggs, assistant to Johnson and liaison man to the committee, that others would be named. When the committee was struck, the members were advised by Meggs that it was an "open committee," that it could talk to anyone within or without the CBC, but it was advisory in nature only. The expenses were paid by the CBC not the House Broadcasting Committee as many farmers thought.

After some initial close monitoring by top CBC staff, the committee was left on its own to dig into questions about where the farm broadcasters fitted into the CBC organization, why there was no farm chief of staff, whether their budgets were being cut and why so many solid, down-to-earth farm writers, who had been on staff at the CBC a long time and had solid followings, had be rousted out by various bureaucratic techniques and program format changes and, if replaced at all, were replaced disproportionately by women.

Reg Forbes of Brandon, former president of the Agricultural Institute of Canada, was chosen to head the committee. It met only four times in 1979 in Vancouver to St. John's and points in between. Allen and others found they were coming up against a dead end even in such simple matters as obtaining a CBC organizational chart—because the CBC couldn't produce one.

"And I still don't know how the money end of it works," said Allen. Staff hiring and firing is done by committee. "When we met in St. John's in July of 1979, we learned that an internal committee had been set up to examine the English agricultural television. Its chief recommendation was to fire Don Baron, head of the department, because some of the Western wheat pools and farm organizations didn't like his philosophical approach to agriculture and had gotten to Johnson about him," said Allen. "The committee had been disbanded and fired Baron before we even knew of its existence. He was replaced by a man who knew little about agriculture. Baron was presenting the real issues faced by agriculture in prime time. The CBC wanted him to push its brand of politics in these presentations but he didn't. Forbes hit the roof and demanded to know what his committee was needed for when it wasn't even consulted in such an important matter. He was told to cool it by the new man."

The new liaison man is Andre Lamy, who was formerly with the National Film Board. Meggs was shuffled off somewhere in the maze. Lamy put on a gag rule, which had the effect of making the members a tame committee. In response to a query by John Wise, Conservative minister of agriculture, as to what was going on, a member of the CBC farm staff indicated Forbes "was a second-rater and not the CBC choice of a chairman."

And recalling that Margaret Lyons, head of the English radio network, had referred to CBC farm broadcasting as ghetto broadcasting—a type of broadcasting that should be eliminated—Allen came to the conclusion he was wasting his time on the Forbes committee. Despite entreaties by Mrs. Lyons to stay, he left for good. So brazenly fine-tuned is the CBC bureaucracy that he received a call from Gerry Wade, a CBC Toronto farm writer, asking him to name a candidate for his successor. "I thought this was a move contrived by the CBC so it could say I recommended my own successor and didn't resign in protest," Allen concluded.

Allen's form of grass-root protest may be what is needed to put CBC farm broadcasting on a higher priority in the CBC.

"DRAGONLADY" HAS "DRAGONMAN" WATCHING — *The Calgary Herald* - September 12, 1984

The ability of human beings to handle power once power has been thrust upon them or once they have grasped power is an ongoing source of amazement. The Alberta Women's Week at Olds College provided a mini-study in the use of power. It all happened because Margaret Lyons, vice-president of English radio for the Canadian Broadcasting Corp., came to speak.

Lyons and I have two things in common. We are both 61 years of age. But whereas she tells me my looks belie my age, I can also say the same of her. She could still pass for a round-faced college sophomore but it is many years since she emerged from McMaster University in Hamilton, Ontario with a degree in economics.

The fact that she was a member of a Canadian family of Japanese extraction who was moved out of a small farm at Mission City in the Lower Mainland of British Columbia during the Second World War was a source of fascination to many of the farm women at Olds. Following her formal speech, she was asked several questions about her feelings today. She gave a short account of the forced exodus inland of her family—a journey that resulted in eventual relocation in Winnipeg. While the men were sent out to hoe sugar beets and other menial farm tasks, she and her sister did domestic work for a family in the wealthy Tuxedo area of Winnipeg. Her comment on the whole traumatic experience was that while first-generation Japanese-Canadians numbly accepted their wartime fate, the second and third generations in the mainstream of Canadian life are beginning to raise protests and demand financial restitution.

To her, the "dragonman" in the lives of her people at that time was the enemy alien property custodian. This individual—probably a plodding civil servant who had the reins of power thrust into his hands—was given authority to seize the property of the evacuees and sell it. Naturally, these distress sales, in effect, dispossessed this group of Canadians. Despite the fact it was a wartime act "in the national interest" and done out of fear, the injustices brought about by this dragonman still irk Margaret Lyons every time she thinks about them. However, a career as a wife, mother and broadcast executive doesn't give her much time to think about them today.

As a result of aid available through the circumstances of living in the Tuxedo district, she was able to go forth and acquire that university education—and took off for England. One day she walked in the front door of the British Broadcasting Corp. and asked for a job as a producer. But she didn't have a curriculum vitae that suited the BBC and was shown the door. Then, she explained to the women at Olds, she got into the BBC via the back door by obtaining a job in the typing pool. Being Canadian, she was later assigned a job in the French newsroom (all Canadians abroad are regarded as French). She was able to assist the French-speaking Algerian desk with her knowledge of economics. Then she went into the Japanese-language section and later became a producer. The BBC found she was a very bright person.

A chance interview of Lester B. Pearson, then Canadian prime minister, brought her back to Canada to work with CBC radio in 1960—after eight years with the BBC. At that time television was just coming on stream and AM radio was in the doldrums as millions were apparently deserting it. Although the programming was topnotch, the task assigned to CBC radio management was redirecting energies to keep it dynamic. "They were getting ready to abandon radio for this new television. It was a close call," she said. The redirection of energy took the form of carving up the daytime schedules to put local programming into prime time national broadcasts and introducing a better type of news and public affairs broadcasting, for example: As It Happens.

A great many well-loved skilled broadcasters and freelancers found themselves shoved off into limbo or out of radio in this "new wave." There was uncertainty and dispossession—and Lyons became known in the studios across the country as the "dragonlady" from Toronto. I first heard this description of her from my friends in the farm broadcast department who were suffering radical changes. Lyons and others in the management team wield power over many people—and continue to do so because they believe it is good for the CBC and the nation. Now as Lyons thinks of the injustice wrought by her dragonman in far-off British Columbia, there are those who think of her as the dragonlady in far-off Toronto, who has deprived them of their heritage as Canadian broadcasters.

She and her fellows have an "out," however. They had another dragonman looking over their shoulders until he died last November. He was Graham Spry, head of the Canadian Radio League, who constantly agitated for topflight programming. Now that he's gone, there is only Dave Kirk, secretary-manager of the Canadian Federation of Agriculture, to carry on this work.

THREATS OF LAWSUIT FORCE CANCELLATION OF CBC's FOOTAGE — *The Calgary Herald* - November 21, 1984

A curious sequel has occurred in a story that made big headlines during the summer. The story concerned charges by the federal meat inspectors' union that shipments of "contaminated" Canadian beef had been halted by foreign importers, returned to Canada, reworked here and sold to Canadians. This all occurred, it was alleged, because the federal Department of Agriculture had cut down the number and frequency of meat inspections.

The Canadian Cattlemen's Association, backed by the Alberta Cattle Commission, became so concerned about these charges floating around the country that it bundled up a fistful of press clippings, took them to Ottawa, and set up a conference with the offending union and the Agriculture Department's meat inspection veterinarians. There were no big headlines about the outcome of this meeting. But the union backed off on all its charges because it was unable to substantiate any of them. Its members agreed to button their lips under such circumstances in future.

The circumstances were outlined by Jim Graham, of Rainier, ACC president, at a zone election meeting in Strathmore. The 1,200-member meat inspectors' union was having a big election fight, with officers being challenged from the ranks. They needed an "issue." At the same time, the union was being challenged by its employer, the federal Agriculture Department. It seems the government wanted to revise inspectors' working schedules to effect more productivity. Graham said bogus scheduling in certain areas had allowed the members to collect vast amounts in overtime. A big game of politics was being played and, as a third party, the Canadian beef industry was being victimized. This is not the first time this has happened to Canadian farmers on account of labour union disruptions and they are a little sick of it.

Then the union broke the story about a meat shipment from a Canadian packing plant which was refused at an English port because of alleged contamination. It charged lack of inspection was responsible. Graham said the facts of the matter are different than the union's allegations. Between the time the British importer had placed the order and the time it arrived (this was over nine months ago), the price of meat went down and he stood to take a bath on the deal. He looked around for a way to break the contract and he found it. (I can't understand how it was possible to place an order of meat into a European Economic Community country in the first place. The EEC has meat running out its ears. Subsidized beef exports from another EEC country, Ireland, have been plaguing the Canadians cattle industry for months.)

The meat in the Canadian shipment was wrapped in plastic and it was unwrapped for inspection before it was completely thawed out. The plastic naturally tore and left some dirt on the wrapping stuck to the meat. Some of the meat chunks were undersized. So with these two strokes, the importer got himself off the hook by rejecting the shipment as "not as ordered." It was returned to Canada—and was put back in the stream here.

However, before it was admitted through customs, it was required to go through Canadian veterinary inspection, which it passed. Faced with these facts and others which silenced its charges, the union backed off and agreed it had been in the wrong and would make no more public statements that gave a black eye to one of the world's best meat-grading systems and also to the Canadian beef cattlemen who produce some of the world's highest-quality product.

Early in October, the Toronto office of CCA learned through a leak that the Canadian Broadcasting Corp. television show, Marketplace, hosted by Bill Paul and Christine Johnson, had planned to revive the issue of "contaminated" meat on a forthcoming show. It could not be determined whether the CBC approached the meat inspectors union or vice-versa but there was certainly no input from the cattlemen nor was any reference made to the fact the CCA had knocked out the union on every charge it had made. Provoked into action on this further intervention, the CCA let the CBC know it intended to take legal action to force them to keep their footage in the can. After hearing the cattlemen's position, the CBC powers-that-be pulled this segment out of the schedule of Marketplace.

"It is for this reason some of us farmers don't have much sympathy for the CBC when the government decides to knock $85 million out of its budget," said Graham. However, there is a certain irresponsible lot in the CBC, which have done a lot of things not in the best interests of this nation and its farmers. "We appreciate the intelligent coverage of agricultural issues CBC farm writers have given the public over the years," Graham added. Perhaps that $85-million cut handed out by Finance Minister Michael Wilson is a signal to some of the other lot that the public is getting fed up with them—not necessarily the CBC itself.

MPs SHY FROM ISSUES NOT COVERED BY TELEVISION
— *The Calgary Herald* - January 24, 1986

Was the last federal election held merely to eradicate the unlamented Pierre Trudeau from national politics? And has he been replaced today by government action developed by television cameras focused on pressure points? Do the Members of Parliament disregard important issues which are too complicated for the television cameras to handle or which their operators choose to ignore because of lack of pizzazz? My mind keeps going back to a speech given by Lee Clark, MP for Brandon-Souris and chairman of the House of Commons agriculture committee, to a farm writer group. He made the point that "we (the Conservatives) are losing headlines by fine-tuning existing programs to assist agriculture."

What he was trying to get over was that the agricultural policy of the Conservatives is to work with the grassroots and then to assist the grassroots by either fine-tuning present agricultural legislation to make it work effectively or replace it with better programs. However, in developing such programs the majority party was at a disadvantage because the members were outside the range of the television crews and Ottawa reporters. "For years farmers and farm leaders have accused the federal government of bringing forth ad hoc agricultural programs. But when we take the time to try to think things out and try to develop long-range programs, we are constantly interrupted by the television play on minor issues," Clark said.

A busload of farmers dogging the agriculture minister here, a few tobacco growers there in a standoff with police on the Hill, a Saskatchewan farmer in the middle of a dried-out field or a tractor parade about some regional grievance will yield two minutes on the night news. The issue may not mean all that much to some of those who were photographed in a protest action but they all go home at night to see themselves on television—just like the Irish Revolutionary Army after its members shoot a couple of British soldiers.

"Therefore," said Clark, "when the television cameras discovered the Western drought last year, those farmers got an infusion of government funds." The scary part of all this is that it is only human nature to find that MPs are confining their activities to dealing with issues that get them television footage. Why spend time on committees away in back rooms working on real issues in agriculture when a 10-second clip on some innocuous fluffy issue will make the National or the Global News? Equally frightening is that unless the issues of agriculture can be brought down to the level of a Moammar Khadafy press conference threatening to blow up the Middle East or pictures of Canadians lining up to buy lottery tickets, agricultue is not likely to be touched.

Clark and his fellow Conservatives have ruefully come to the conclusion the size of their election victory was counter-productive. What it amounted to was the people just wanted a different prime minister to announce government handouts. The electorate had incredible expectations of the Tories as they waited so long for the change in government. But the opposition doesn't need any brains or statesmanship any more. It just needs a television camera and a placard writer.

NDP DEREK FOX IS IMPATIENT WITH PRESS GALLERY
— *The Calgary Herald* - February 12, 1992

Derek Fox, NDP, MLA for Vegreville, the party's agriculture critic, is upset with the press gallery in the Alberta Legislature at Edmonton. This is an ironic situation. Most press gallery members are anti-conservative. If they don't carry an NDP membership card it's because they are broke like most reporters have been since Moses was a linotype operator and can't afford one. The NDP is concerned with the way the Alberta government is moving in agriculture. But Fox can't get anyone in the press gallery interested in these important economic issues. Says he: "The gallery is preoccupied: too busy concentrating on the latest relative of the premier to be appointed to a government board or the most recent $100 million Conservative fiasco or scandal. They don't put much time into covering agricultural issues."

Fox was so distraught about this regressive tendency that he took to the rural areas himself last summer to explain his party's position on agriculture to anyone who would listen. And speaking about people listening, Fox claims practically nobody can catch Agriculture Minister Ernie Isley's ear any more. "Even a mostly sympathetic group like Unifarm passed a motion saying it cannot talk to Isley about issues any more. He doesn't listen. Unifarm wants to bypass him and lobby the Conservative caucus directly," he said. "Isley reflects the common thread that the government doesn't hear what the people have to say—or care. Governments begin to think they know what is best for people. They don't listen or seek input from people. They just do things to people rather than for them. The symptoms: they become increasingly aloof, secretive, arrogant, mean-spirited and preoccupied with maintaining power."

Another reason Fox is so much upset about not gaining the ear of the press gallery is that most NDP members are in the city ridings. Only three are in the rural areas. He wants the city voters to gain an appreciation of the waste in the agriculture department. At a time when Isley announced downsizing his department and a reduction in the number of district agriculturists, Fox suggested the government reduce the number of agriculture ministers to one from two. Both Isley and associate Shirley McClellan have a duplication of perks and bureaucrats. Cutting out one would save $250,000. "It is debatable which minister should be kept," said Fox.

This should be an easy problem to resolve. Leaving should be the minister who kept his department in a state of turmoil for a year and a half by announcing 250 agriculture department personnel in Edmonton were going to the country—but casually cancelled the move just before Christmas.

The real reason the press gallery won't write much agriculture copy is that it doesn't fit into the mold of advanced editorial preparation most big newsrooms indulge in. This is a system whereby news agendas are planned for a week, two weeks or a month ahead. This kind of journalism has no time nor space for Fox and his agricultural problems, in any particular area, except game ranching (which happens to be on the press gallery agenda for some obscure reason or other). The NDP isn't against game ranching. Its position is that before a full-fledged game-ranching industry is established in this province, its proponents shouldn't be afraid of an environmental assessment. Such an assessment might surprise everyone including Fox. If things are as grim as his party alleges, agricultural diversification becomes more attractive. Game ranching is one neat means of diversification. If people won't eat beef, feed 'em venison.

NEWS ANCHOR FAILS TO ATTRACT CBC CAMERAS — *The Calgary Herald* - May 5, 1992

The Agriculture Institute of Canada picked a well-known personage to deliver the high-profile Klinck Lecture for 1992. He is known to millions as the suave, popular, and friendly weekend anchor of the CBC TV flagship news program, The National, Knowlton Nash. The lecture saw him reverse his role to newsmaker from newsreader. Members who seldom show up at meetings turned out to hear him speak at 24 AIC branches across Canada. The Klinck Lecture committee of AIC made an inspired choice when they brought Nash aboard. He has a background in agriculture. He had a chance to view agriculture from a world-wide perspective from 1951-56 in Washington. He was director of information for the International Federation of Agriculture Producers, of which the Canadian Federation of Agriculture is an affiliate.

When he joined the CBC full time in 1961, he became one of the news and public affairs power brokers. That fact inspired some of the 8,000 farm "professionals" (AIC members all have agriculture degrees from universities) to quiz Nash on CBC policy. The CBC has not been a staunch defender and reporter of the agricultural scene in recent years. It has become a consumer and environmentally-orientated network which has pretty well retreated into the never-never of the inner city and sent agriculture out to pasture. Here was a chance for the embattled agriculturists to find out why.

In Calgary, AIC members are well-mannered and polite. They sat through the lecture, the broad outlines of which were that if Canadian agriculture is to avoid economic disintegration, it has to do a lot more in generating broad public awareness of the economics of farming in Canada and the role of agriculture in Canadian society. Nash alluded to the fact the CBC once had a top stable of farm writers. In fact, the CBC built its reputation as a national unifying force in Canada around farm writers like Orville Shugg, George Atkins, Bob Knowles and Norn Garriock. Lately not only the CBC but the big urban media have decided they can't afford to have farm writers on staff; they are in the nature of a crutch similar to agriculture. Therefore, the farmers and farm organizations are left to their own resources for bringing their point of view onto the national and world stage in competition with hundreds of other sexier pressure groups grabbing for space and time.

At the end of Nash's lecture, Frank Jacobs, the doyen of Canadian farm writers, rose and deftly pointed out how very difficult it is for farm groups to get any space or time in the media, faced with this considerable hurdle. "You, Mr. Nash, a high-profile, beloved Canadian newsmaker have been invited into the inner city of Calgary to deliver a high-profile, incisive thoughtful lecture on Canadian agriculture (one of Canada's basic industries). But I do not see one TV crew or urban newspaper writer here tonight to cover your appearance."

Nash knew that Jacobs was adding his voice to the critics of the CBC for not replacing George Price as its agriculture specialist broadcasting from Ottawa when he retired a couple of years ago. This has really bugged the agricultural community. In answer, Nash merely reiterated the main point of his speech:

farmers would have to go it alone—and that Price's job had been a victim of budget cuts. I have it on good authority that Nash, having heard this same message right across the nation, took it back to Toronto to the CBC brass. They are now trying to revive the position and are looking for someone to do it.

Another question from the floor caught Nash somewhat off-guard. This writer asked if the CBC's "creative renewal" stratagem could be applied in the agriculture community to assist it to do an improved job of reaching the general public. Creative renewal is the CBC's long-range strategy to "position" its services in the '90s as a means of fending off a body of Members of Parliament who would like to see it deregulated or privatized, thus putting it at the mercy of advertisers. Nash said the short answer to this is "no" and he admitted some of the stratagems dreamed up by senior management are driving professional broadcasters crazy. CBC aficionados are also having difficulty coping with wholesale trivialization of the top-rated programming. From liberal arts programming, the CBC deliberately regressed to jungle music and talk radio to satisfy the dead-end yuppie crowd. Perhaps the CBC supremos in their haste to give Canadians formula broadcasting have forgotten the farm broadcasts used to win top ratings.

Schmidt is pleased that he has been able to win so many friends and influence so many people despite the fact that he has never read Dale Carnegie's book on how to go about doing just that. Farmers today are still dealing with: new ideas, imaginative perspectives on old problems, old failures that are repeated, and regressive strictures which are taken against the farmers left in agriculture. He believes that card-carrying farm writers can still contribute to the bona fide farmers and public a greater understanding of these recurring matters.

EPILOGUE

Some thirty years ago, John T. Schmidt wrote:

> **"History as written by the professional historians can be deadly dull and stuffy and full of dates and places that never bring out any human interest. And what is history other than a great number of human experiences told by the people who were actually on the scene at the time?"**

This biographical profile of the author of these words has made every attempt to adhere to the wishes he expressed on August 20, 1970. Rather than fill these pages with "deadly dull and stuffy" information about "dates and places that never bring out any human interest," it enables Schmidt's words to speak for themselves about some of the people and experiences he had when he was "actually on the scene at the time." Through this random collection of his newspaper articles, we come to understand his secret for a long, healthy and extraordinarily productive life. Moreover, we can apply his secret to our own lives by adopting his answers to two key questions that are pivotal to this book:

1. What keeps a person doing what he has been doing for more than sixty years?
   Schmidt's answer is—do what you really love to do.

Certainly, we all need to earn a living but his secret for longevity is to find something that you really enjoy and find a way to keep doing it. Even though he no longer commands the kind of monetary compensation that he did while working for a large daily newspaper, he continues to write as though he did. Moreover, he believes himself to be richly rewarded by the sheer enjoyment of his enduring fascination for people and the world. Interestingly, his answer to the first question is directly connected to the second:

2. What keeps people reading the work of a writer who has written more than 10,000 newspaper and magazine columns and several books?
   Schmidt's answer—his readers share his insatiable curiosity.

Reflected in his writing, his personal life's philosophy is to avoid being self-consumed. Nothing is more destructive to the human spirit than being obsessive about oneself. Schmidt maintains that true happiness is found by taking personal responsibility for making this world better for having been in it. To enjoy a long and happy life—think of yourself within the larger context of the big, beautiful world around us. And above all, keep your worries in perspective with a bit of humour by remembering that most problems don't even qualify as Quality BullSchmidt.

SCHMIDT'S NEWSPAPER CONNECTIONS

Few, if any Canadian families in the newspaper industry, can present a lineage like Schmidt.

THOMAS MUIR ANDERSON
(Great-Grandfather)
Financial writer in 1860s for the Toronto Globe, owned by George Brown a Father of Confederation.

JOHN B. SCHMIDT
(Great-Uncle)
Fought with Governor General's Footguard in Riel Rebellion. Employed in composing room of Toronto Mail in 1880s.

THE SCHMIDT BROTHERS — John T. (left) with brothers James W. (centre), and Robert A. (right) and family dog "Sandy" at home in Ayr, Ont. All were brought into the newspaper business by father John A. who purchased the Ayr News in 1913. Robert passed away in 1968. James, like John T., has been in newspapering for over 60 years, and today (2003) continues as head of the independent, family-owned, News.

JOHN A. SCHMIDT (Father)
Purchased Ayr News 1913

JOHN T. SCHMIDT

JOHN P. SCHMIDT (Son)
Editor Ayr News 2003

THE EARLY YEARS . . .

A young John T. Schmidt.

An early venture into farming.

Family pet chicken "Nummy" had run of the house.

Ayr School where John T. received both elementary and secondary education.

AT WORK AND LEISURE . . .

Praying for better mail service which collapsed when post office quit using passenger trains to carry mail.

John T. in sanctum sanctorum.

Eating national dish of Waterloo County . . . 1 pigtail, 1 sauerkraut, 1 scoop mashed potatoes, 1 Lamb's Navy Rum and Coke.

Canada's No. 1 computer hater John T. hacking viruses to pieces in a fund-raising event by the Drumheller Library.

John T. and wife Margaret travel by limo to his 65th birthday party.

Riding the rails is the favoured mode of transportation. Will ride in engine if necessary, thus the cap.

Discussing politics with Del Pound (left) ex-Chief Commissioner of Canadian Grain Commission.

Checking out the gossip at a bull sale.

This is no bull — John T. could get around by horse and buggy when necessary.